William D. Granger

How to Care for the Insane

A Manual for Nurses

William D. Granger

How to Care for the Insane
A Manual for Nurses

ISBN/EAN: 9783337372750

Printed in Europe, USA, Canada, Australia, Japan

Cover: Foto ©berggeist007 / pixelio.de

More available books at **www.hansebooks.com**

HOW TO CARE FOR THE INSANE

A MANUAL FOR NURSES

BY

WILLIAM D. GRANGER, M.D.

PROPRIETOR-PHYSICIAN, VERNON HOUSE, MT. VERNON, N. Y.

FORMERLY FIRST ASSISTANT PHYSICIAN BUFFALO STATE HOSPITAL, BUFFALO, N. Y.
MEMBER AMERICAN ASSOCIATION OF SUPERINTENDENTS OF HOSPITALS FOR
THE INSANE. MEMBER NEW YORK NEUROLOGICAL SOCIETY.

SECOND EDITION. REVISED.

G. P. PUTNAM'S SONS
NEW YORK　　　　LONDON
27 WEST TWENTY-THIRD ST.　27 KING WILLIAM ST., STRAND

The Knickerbocker Press

1891

NOTE TO SECOND EDITION.

AT the time of starting a training school, in 1883, the author was unaware that like work was commencing at the McLean Asylum, Somerville, Mass., by Dr. Campbell Clark and others in Scotland, and in New South Wales.

Each was independent, and each worked out the problem independently. Thus, far separated efforts showed the time had come when attendants must be evolved into trained nurses.

The most gratifying feature has been the unanimous approval by American superintendents and the establishment of training schools in almost every asylum in the land, often under the most difficult conditions and at a great sacrifice of the precious time and strength of the medical staff.

The increasing number of these schools and pupils demands a second edition of this little manual.

VERNON HOUSE, MT. VERNON, N. Y.,
    March 21, 1891.

INTRODUCTION.

THE writer began in October, 1883, at the Buffalo State Asylum for the Insane, a course of instruction to the women attendants upon their duties and how best to care for their patients. This has been regularly continued till it has become a fixed part of the asylum life, and has developed into a system of training, and now a class of attendants has nearly completed its studies. Since July, 1885, instruction has been given to men attendants.

In April, 1885, the Superintendent, Dr. J. B. Andrews, who had encouraged the school from its conception, asked the Board of Managers to officially recognize it. They adopted the recommendation and fixed the qualifications for admission, the pay and privileges of its members, and provided for a certificate as a trained nurse and an attendant upon the insane, to be given to all, who at the end of two years successfully finished the full course of instruction.

The writer believes that all attendants should be regularly instructed in their duties, and the highest standard of care can be reached only when this is done. He also believes that every person who is allowed to care for the insane will be greatly benefited by such instruction, and will be able to learn every thing taught, if the teacher uses simple methods and is patient to instruct.

As a rule they enter upon the study with interest, and soon a skilled corps is formed, who are competent to fill the responsible positions, and control the unstable class that drift in and out of an asylum. Even the dullest are awakened to new zeal, and are advanced to positions of trust they could not otherwise have filled.

A brief outline of the course of instruction of the school may be of interest.

The first year is spent in learning the routine of ward work and filling minor positions. The attendants are changed from ward to ward, and have the care of all classes of the insane.

They first receive instruction in the printed rules of the asylum. Every rule relating to the duties of attendants is read and explained, and special attention is called to the performance of the following duties :

a. Duties to officers.
b. Duties to each other.
c. Duties to patients.
d. Duties to the institution.

Thus the new attendants early get an outline of their duties in the special care of the insane.

After this comes instruction in elementary anatomy and physiology. They are taught of the bones, joints, muscles, and organs of the body, food and digestion, the circulation and respiration, waste and repair, animal heat, and the nervous system.

In order to be ready for advanced instruction the elements of physiology must be thoroughly learned. The teaching must be adapted to the ability and wants of those instructed. Having fixed the limit of duties required of an attendant, it is easy to fix the limit of instruction. It is an error to teach too much medicine, for then we begin to make physicians. All that is needed is attendants who are able to do their work intelligently, and, keeping this object in mind, lectures by a physician, devoid of too much detail, but simple, direct, and plain, are better than instruction from any of the text-books. With notes of the lectures furnished, and with repeated recitations, any lesson is readily learned. This way of instructing, by lectures, notes, and recitations, is continued throughout the entire two years.

A course in hygiene follows the lectures in physiology.

Instruction in these three studies occupies the first year. An attendant who, at the end of this time, successfully passes an examination in them, and who has been faithful in his duties, is ready to receive the advanced instruction of the second year. This includes the nursing of the sick, the management of emergencies, and finally the special work of caring for the insane. The wits of an attendant upon the insane have to be sharpened in many directions not required of a general nurse. The textbooks on nursing may properly be followed by another, which shall aid one skilled as a nurse to perform the varied and difficult duties incident to the care of the insane and the wards of an asylum. To furnish this is the object of this manual.

A brief review of the physiology of the nervous system is introduced for the aid of students, in reading the chapters on the mind and insanity.

To teach any thing metaphysical or pathological may seem questionable. The class, however, has not only been interested in the simple study of the phenomena of the mind, but has been able to comprehend and profit by the lectures on this subject.

The lectures on the care of the insane were given to the class almost as they appear in these pages. The suggestion was made that if they were printed they would find a place in the hands of attendants in other asylums. This is the reason of their publication.

To my colleague, Dr. A. W. Hurd, I wish to tender my thanks for the valuable assistance he has given me in the preparation of this manual. I am greatly indebted to Dr. Andrews for his ever kind but critical advice. But for his encouragement and help neither the work of instruction nor the preparation of these pages would have been begun, nor success, if success be gained, achieved.

CONTENTS.

	PAGE
INTRODUCTION	v

CHAPTER I.

THE NERVOUS SYSTEM AND SOME OF ITS MORE IMPORTANT FUNCTIONS   1

Nerve Centres.—Brain and Spinal Cord.—The Nerves.—Nerve Cells and Fibres.—Motor and Sensory Nerves.—The Five Organs of Special Sense.—Nerve Impulses.—The Brain and Nervous System Always Busy.—Need of Rest.

CHAPTER II.

THE MIND AND SOME OF ITS FACULTIES   7

Mind and Matter.—Life.—Relation of Mind and Brain.—Faculties of the Mind.—Intellectual Faculties.—Will.—Emotions.—Instincts.—Moral Faculties.

CHAPTER III.

INSANITY; OR, DISEASE OF THE MIND   13

Insanity a Change. — Involves Disease of the Brain.—Delusions. — Hallucinations. — Illusions.—Incoherence. — Mental States. — Mania. — Melancholia. — Dementia. — Monomania.—Emotional Insanity.—Dipsomania.—Moral Insanity.

CONTENTS.

CHAPTER IV.

THE DUTIES OF AN ATTENDANT 22

What an Attendant Should First Learn.—The Relation of Attendants to Patients.—The Character of an Attendant.—Relation to the Institution.—How and What to Observe.—Systematized Plan of Observation.—Control and Influence of Attendants over Patients.—Care and Study of the Individual.—Liberty to be Allowed Patients.—Self-Control of Patients to be Encouraged.

CHAPTER V.

GENERAL CARE OF THE INSANE 33

Reception of New Patients.—Work and Employment.—Patients' Care of Themselves.—Walking.—Clothing.—Bathing.—Serving of Food.—Bed and Rising Time.—Night Care.

CHAPTER VI.

CARE OF THE VIOLENT INSANE 45

Need of Studying Each Case.—Constant Attention and Oversight.—Value of Employment and Out-Door Exercise.—Restriction and Idleness.—Paroxysms of Violence; How Cared For.—How to Hold or Carry a Patient.—Danger of Injury.—Struggles to be Avoided.—Care of Destructive Patients.—Use of Restraint, Seclusion, and Covered Bed.

CHAPTER VII.

CARE OF THE HOMICIDAL AND SUICIDAL INSANE, AND OF THOSE INCLINED TO ACTS OF VIOLENCE . . . 53

Delusions of Suspicion.—Homicidal Patients.—Suicidal Patients.—Self-Mutilation.—Incendiary Patients.

CHAPTER VIII.

CARE OF SOME OF THE COMMON MENTAL STATES AND THE ACCOMPANYING BODILY CONDITIONS 60

Care in the Earlier Stages.—Insanity with Exhaustion.—Symptoms of Danger.—Care of Dementia, Early Dementia, Chronic or Terminal Dementia.—Convalescence.—Relapse.—Epilepsy.—Paresis.—Care of Paralytics, the Helpless, the Bed-ridden.—Bed-Sores.

CHAPTER IX.

SOME OF THE COMMON ACCIDENTS AMONG THE INSANE, AND THE TREATMENT OF EMERGENCIES 71

Certain Classes of Insane Liable to Injury.—Fractures.—Wounds.—Bites.—Blows on the Head—Cut Throat.—Wounds of the Extremities with Hemorrhage.—Sprains.—Choking.—Artificial Respiration.—Burns.—Frost-bites.—States of Unconsciousness. — Apoplexy. — Sunstroke. — Poisoning.—Eating Glass.—Injury with Needles.

CHAPTER X.

SOME SERVICES FREQUENTLY DEMANDED OF ATTENDANTS AND HOW TO DO THEM 85

Administration and Effects of Medicine.—Opium, Chloral, Hyoscine, and Hyoscyamine; Doses, Effects, Poisoning, Treatment.—Stimulants.—Applications of Heat and Cold.—Baths and Wet Packing.—Hypodermic Injections.—Forcible Feeding with Stomach-Tube.—Nutritive Enemata.

HOW TO CARE FOR THE INSANE.

CHAPTER I.

THE NERVOUS SYSTEM AND SOME OF ITS MORE IMPORTANT FUNCTIONS.

THE nervous system is made up of a nerve centre and nerves.

The great nerve centre is the *Brain* and *Spinal Cord*.

The brain is a body weighing about forty ounces, and fills a cavity in the upper part of the skull. The spinal cord, commonly called spinal marrow, is directly connected with the brain. The skull rests upon the spinal column, or backbone, and there is a cavity inside the whole length of this column, which contains the cord. There is an opening through the base of the skull where it rests upon the spinal column, and it is through this opening that the fibres of the cord go, to pass into and become a part of the brain. These most important parts are carefully protected by a strong bony covering.

Many nerves are given off from the brain and cord and go practically everywhere, so that every part of the body is supplied with them. These nerves are white cords of different sizes; the largest nerve of the body, the one that goes to the leg, called the sciatic, is as large as the little finger.

There are really two brains and two cords, as along the central line of the body there is a division of the brain and cord, making two halves exactly alike. These halves are connected together, the division not being complete.

Nerves are given off in pairs; for example, from either side of the brain arises a nerve that goes to each eye. So two nerves exactly alike spring from the two sides of the spinal cord, going to each arm.

A nerve is composed of a bundle of fibres, microscopic in size. As a nerve passes to the extremities it divides by branching much as does an artery, and thus a bundle of fibres is distributed to a muscle, or a part of the skin, or to an organ, and every part of the body has a direct nerve supply, much as you saw in the microscope it was supplied with blood by means of the capillaries. We cannot prick our finger with the finest needle but nerve fibres are irritated, and we feel it, and capillaries are injured and we get a drop of blood.

Most of the nerves that go to the arms, legs, and organs of the chest and abdomen, arise in and proceed from the spinal cord, but some of the fibres begin in the brain and are continued down the cord, where, joining with fibres that originate in the cord itself, both go to make up the nerve, thus connecting all parts of the body with the great centre.

The brain and cord are made up of blood-vessels, nerve cells, nerve fibres, and, holding them all together, connective tissue. The cells are very small, being microscopic in size; there are an immense number of them, and they make up most of the gray matter or outside of the brain, but in the spinal cord the gray matter is in the centre. The fibres that go to make up the nerves begin and spring from the cells, and they also unite them together.

The cells are gathered into groups, which have each a separate function to perform. There is a group from which the nerve of the eye proceeds; another for the nerve that goes to the ear; another for the nerve that goes to the arm; and another for the nerve of the heart. There is a group that presides over speech, and other groups that preside over mental action, while all of these are connected together by fibres. Thus it appears that the brain is a true "centre," and the nerves but the means of connection between different parts of the body and the brain, and also between different parts of the brain.

Nerves have two special functions: one to carry impressions made upon the fibres, that end in the different parts and organs of the body, to the brain; another to carry from the nerve cells so-called "nerve impulses," to the different parts and organs of the body. Some nerves have in themselves these two functions, as the nerves that go to the arm or leg; others have but one, as the optic or eye nerve, which can only carry the sensation of sight from the eye to the brain.

The nerves that carry sensations to the brain are called *Sensory Nerves*. The nerves that carry motor impulses from the brain are called *Motor Nerves*.

There are five special organs of sense, each receiving different impressions, and sending by its sensory nerve or nerves a different character of sensation to the brain, namely :

The eye, giving sensations of light and color.

The ear, giving sensations of sound.

The nose, giving sensations of smell.

The mouth, giving sensations of taste.

The skin, giving sensations of touch, with ideas of roughness, smoothness, hardness, softness, heat, and cold.

There must be, in every case, a direct nerve connection from the organ of special sense to the special group of cells in the brain to which the nerve goes. If the connection is broken at any point, the impression made upon the fibres in the organ of sense cannot reach the brain. Only after the impression reaches the brain and the cells are affected, do we become conscious of a sensation. We then say, as the case may be, I see, or hear, or smell, or taste, or feel something.

It thus appears that these organs of sense simply receive the impressions made upon them to transmit to the brain, and it is really the brain that sees, hears, smells, tastes, and feels. By the action of the organs and nerves of special sense we get all our knowledge of the external world, and, probably, if we had no organs of sense, we would have no consciousness of our existence.

Pain is due to abnormal action of sensory nerves, caused by disease, injury, or pressure, and the irritation made, being carried to the brain makes us conscious of the peculiar sensation we call pain. So the want of food or water makes an impression upon nerves, which being carried to the brain causes a peculiar sensation, and we say we feel hungry or thirsty.

The *Motor Nerves* arise in the cells of the brain and cord. Those which go to the voluntary muscles cause them to contract, and are under the control of the will. If the cells are diseased, if they do not get enough arterial blood, or are poisoned by carbonic acid, or if the nerves are diseased, injured, or cut, so that nerve impulses cannot be sent from the brain to the muscles, we have paralysis of a muscle or a group of muscles, according to the extent of the injury. Now we can appreciate the force of this teaching in the physiology of the muscular system, that " paralysis is a loss of power, either partial or complete, to contract muscles, due to disease of the nerves."

By the ready action of our mind, the quick working of our will, we direct and control the action of our muscles, so as to perform with the utmost skill and ease the varied and innumerable movements of our body.

It seems very easy to do this, but watch a child learning to walk ; it is educating its mind and will to control the muscles, and it is a slow and difficult education.

But all motor impulses and bodily activities are not under the control of the will. The heart is supplied with motor nerves, but we cannot by our will stop its beating or control its action. The taking of food makes a mental impression, and without the will being involved, impulses are sent to the glands of the mouth, setting them actively at work, and saliva flows. So the stomach begins to churn food when it is introduced, and the liver is kept at work making bile and sugar, and we breathe when we are asleep.

All the organs of the body are supplied with motor

nerves, that regulate their action and give them the power to do their function or work, but with the exception of the muscles, this power is sent without the action of the will.

Our brains are very busy. While we are awake we are constantly receiving sensations, we are thinking, remembering, willing, and sending many messages every minute, and directing power to all parts of the body. The brain works and gets tired, just as the rest of the body gets tired, and, if abused, injured, or overworked, may become diseased as may any part. Its tissues wear out, are burned up, and require the same supply of material to repair them that any other part of the body requires. It needs then rest, good food, good blood, and plenty of oxygen.

No wonder some brains give out, and fail to do their work properly, and so cause insanity.

CHAPTER II.

THE MIND AND SOME OF ITS FACULTIES.

WE know there is something we call mind, because we know something of its way of working, or its faculties. What mind is we do not know, but we know it is not matter, because matter is something that occupies space, and has qualities that do not belong to mind. We say of mind, it reasons, remembers, or wills ; of matter, that it is hard or soft, or cold or elastic, or that it has color ; speaking always of the faculties of mind or what it does, and of the qualities of matter, or what it is. We do not know what matter is, only how it appears to us ; we know it is not mind because mind is something spiritual, and possessed of faculties or powers that do not belong to matter.

Mind and Matter are the only forms of existence of which we have any knowledge.

We speak of matter as inorganic—that is, without life, as iron, water, oxygen ; and as organic, or matter plus something we call life. Life appears in two forms, namely, vegetable and animal. The lowest forms of animal life have no nervous system, but as we ascend in the scale the nervous system appears, and becomes more and more complete.

Man possesses the most perfect nervous system, has

the most perfect brain, and also an intelligence far above that of any other animal, and is endowed with some mental faculties that belong to him alone. The brain may be said to be the organ of the mind, but we do not know what is the true relation between them; that is, how the brain is acted upon by the mind, or how the action of the mind affects the brain. Brain is matter, and very solid matter as well, mind is immaterial, or spiritual, and the exact connection between something material and something spiritual has never been made out and never will be.

Some say the brain makes mind a good deal as liver makes bile, or the glands of the mouth make saliva, or the cells of the brain make motor impulses, and if the brain does not act there is no mind made; so much cell action, so much memory, reason, or will produced. But how, it is immediately asked, is something material to make something immaterial? Others say that mind is something, and has an existence of its own, and, though spiritual, acts upon its organ, the brain, and by so doing, we are conscious that we see, reason, remember, and will. But how, it is immediately asked, does something immaterial act upon something material? We do not know, and we probably never shall know. This intimate connection between mind and matter exists during life only; it begins with life and ends with life.

We must then come back to the starting-point—there are two forms of existence, mind and matter. We do not know what either really is, but only the faculties or working of our minds, and the qualities or appearance of matter.

THE MIND AND SOME OF ITS FACULTIES.

Mind thinks or remembers, reasons or wills, but these are faculties of the mind; it is what the mind does, not mind itself. Gold is yellow, but yellow is not gold; gold is hard, but hardness is not gold; these are qualities of gold, and not gold itself.

In the study of physiology you found the body divided into many parts, and that these parts had each a separate function or duty to perform. In the study of the mind, we find it has many different faculties or ways of working. We did not study all the functions of the body, so we will not study all the faculties of the mind.

The mind is very complicated in its action, and difficult to understand. Men study it all their lives and are not agreed about some of its simple manifestations, and argue and even contend about their differences. There are, however, some seemingly natural divisions of the faculties of the mind, and a knowledge of these is sufficient for our purposes.

We may say of the mind that it possesses:

a. Intellectual faculties.
b. Will.
c. Emotions or feelings.
d. Instincts.
e. Moral faculties or conscience.

The first three are commonly given as divisions of the mind; the last two are included for convenience of teaching.

The Intellectual Faculties include those powers which in common language are called "mind." A few only will be considered—namely, the perceptive faculty, consciousness, memory, and reason.

The Perceptive Faculty is the power of the mind to perceive or know the sensations brought to the brain by the sensory nerves, from the organs of sense, and the action of this faculty gives us a knowledge of the existence and qualities of matter.

Consciousness is that faculty by which we know we perceive, reason, remember, will, or possess emotions. By its operation we know that we exist, have a mind, and what that mind does.

Memory is that faculty by which we are able to recall to consciousness the knowledge we possess of past events.

Reason is that faculty by which we are able to make use of what we know and to acquire new knowledge. For instance, I know the distance between two places is sixty miles, and I know that the cars, going between the places, travel at the rate of twenty miles an hour, and that they leave at four o'clock. Without reason, I could never of myself, know the two new facts, that it would require three hours to make the journey, and that the arrival will be at seven o'clock. The faculty of reason is one of the most distinctive of the human mind.

The Will.—In consequence of our perceptions, our consciousness, our memory, our reason, we are in a condition to know a good deal of what is about us, and of ourselves, and we desire to bring ourselves into relation with the outside world, and therefore we act. There is a faculty of mind that allows us to choose how to act, and this is called the will, or that faculty of the mind "by which we are capable of choosing." By the action of the will, we direct and control the voluntary muscles and motions of the body, while the action of the mind is also largely under its control.

It may truly be said that unless we are under the compulsion of some physical force, we always choose to do whatever we most wish to do. This liberty of choosing is called "freedom of the will," and because we are free to choose, we are responsible for the consequences of our choice. We say, in common language, a person is responsible for what he does, and both human and divine law holds each to a strict accountability for his conduct, because all are free to choose how they will conduct themselves.

The Emotions or Feelings.—The emotions are joy, love, grief, hatred, anger, jealousy, and other like conditions, and we speak of them as "natural," because they appear without the operation of our intellect or will, and the capacity for them seems to be a part of our existence. They should, however, be under the control of reason and will, and a person who gives way to his feelings, as of jealousy, and murders, is held responsible by human and divine law. But though we control them, we cannot prevent their action, and we must, as long as we live, feel love and joy, be affected by grief, suffer from anger, or be jealous.

Instincts.—These belong largely to our animal nature; our appetites and desires are instincts, and we speak of them as "natural." Children want food and drink before they know what it is they want, and birds in the nest, open their little mouths for the worm their mother brings them.

Appetites indulged in become strong, and are often uncontrolled by the reason and will; as the indulged appetite for liquor. A strong and healthy mind should control the appetites, as we have learned it should control the emotions, and we are justly held responsible for the consequences of an indulged appetite.

Moral Faculties.—There exists in the mind of man a knowledge of right and wrong, and a feeling of obligation to respect the rights of others. We can hardly conceive of a man in his right mind who does not know it is wrong to lie, or steal, or murder. The capacity to know right from wrong is called conscience.

Most people, perhaps all, have a feeling of relation and obligation to a higher moral being than man. The feeling to do right because it is pleasing to a God to whom we are directly responsible, is the foundation of our religious convictions.

The mind is a most complex affair, it is always active, nor is one faculty at work and the rest idle, but many parts are at work at the same time, and act and react upon each other. We may exercise our perceptive faculty, or reason, memory, and will, and be affected by our feelings at the same time. There is with it all a regulating power that coördinates or brings these different actions into harmony, and we get the working of a healthy mind.

CHAPTER III.

INSANITY ; OR, DISEASE OF THE MIND.

IN common language we speak of the mind diseased. This is not strictly true, as it is the brain that is diseased and, in consequence, we get disturbed mental action.

Every person has individual characteristics. As no two faces are alike, so the mind, character, and manner of no two are alike, and it is by the manifestation of these, that each is known.

When a person becomes insane there is always a change from his natural way of thinking, feeling, and acting, due to disease of the brain. Sometimes the change is slight, or concealed by the patient, and is apparent only to near friends, or after a careful examination. Sometimes it is so great as to attract immediate attention, when it may present the features of raving madness, or of the most abject melancholy.

To illustrate this change, we may suppose both a king and a pauper to become insane : there is, of course, a vast difference between them, but the king may be so changed by the disease as to believe that he is a pauper, and himself and his family starving, and he may also wish and even try to work and dig like a laborer to support them ; or a pauper may think himself a king, and try to act like one. Such conditions show a *marked* change in the manner of thinking, feeling, and acting, which involves diseased action of the intellect, the emotions, and the will.

Sometimes the appetites are also changed, or control over them is lost, and sometimes the moral nature is affected as well, sometimes a single faculty of the mind appears more disturbed than do others; it is, however, doubtful, or at least denied, that one faculty can show such disturbed mental action as to indicate insanity, and the rest of the mind appear perfectly healthy and normal. With the changes that have been spoken of, there is generally disturbances of the physical health, and often of a marked character. It must be remembered that mere oddity of appearance or eccentricity of conduct, however marked, if natural, do not of themselves constitute signs of insanity.

Some Mental Symptoms of Insanity.—There are some important mental symptoms which quite generally accompany insanity, and are found either alone or combined in the individual case. These are:

a. Delusions.
b. Hallucinations.
c. Illusions.
d. Incoherence of speech.

Delusions are false beliefs. We think a belief in the religion of Mahomet is a delusion, but not an insane one. Insane delusions arise from disease of the brain, and are a part of those mental changes that appear during its progress. The king, who, under the influence of disease, thinks himself a pauper and that he and his family are starving, and the pauper, who thinks himself a king, with all the wealth and power of one, have each insane delusions.

Some delusions are fleeting and changeable, lasting a few days, weeks, or months, while others are fixed, lasting a lifetime ; some are impossible and beyond rational belief, as when a man thinks himself Queen Victoria, or that his head is made of brass, or that he is dead, and yet sleeps and eats and talks ; other delusions are possible, as when a king thinks himself a pauper, because such a thing may and even has happened, or when a pauper thinks himself a king, because people of very low degree have risen to such a station, but they are very improbable, and we do not expect such things among Americans, much less among our patients. Other delusions are not only possible, but relate to things that may or do happen, or are within the bounds of a rational belief, as that of a person who insists he has a cancer, or that he has committed the unpardonable sin, or that poverty is impending and the poorhouse not far off ; or that of a woman that she has been violated, or that, when her child was sick she so neglected it, that it died. Such beliefs as these are delusions, when they have no other reason for their existence than that they are caused by disease.

Some delusions are called homicidal, suicidal, or dangerous, because they cause a patient to do, or want to do, acts that are dangerous to himself or others, or property.

Hallucinations.—When a patient has hallucinations, he thinks he sees, hears, smells, tastes, or feels something, when there is really nothing to cause the sensations or ideas except diseased action of the brain ; nothing being sent to the brain from any special organ of sense, he really sees, hears, smells, tastes, or feels nothing, it is all imagination, though seemingly very real.

For instance, a person thinks he hears a voice, perhaps that of God, or of some one who is dead, or of an absent friend, or thinks he sees these persons, when there is nothing external to the brain to excite the sensation or give the idea.

Illusions.—When illusions are present, the mind fails to perceive correctly what the eye sees, or the ear hears, or the impressions that are brought to the brain from any of the organs of sense. For instance, a person looks at a row of trees, and they appear to him to be a row of soldiers ; or the whistle of a locomotive may be so changed as to seem to be the voice of God ; or the odor of a rose, burning sulphur ; food may taste like poison, or the hand of a friend feel like a piece of ice or a red-hot iron, and is so believed to be. These are deceptions of the senses.

In insanity, the truth and existence of delusions, hallucinations, and illusions are fully believed in, and the patient cannot be argued out of the belief, however absurd or unreal it may be.

Incoherence of Speech.—When a person is incoherent, he rambles in talk ; there is little connection between different sentences, or the sentence itself is meaningless, being a mere jumble of words ; sometimes ideas come too rapidly into the mind, and some new subject is begun and talked about before the first is finished ; sometimes the mind is slow, and memory forgets what is being talked about.

General States of Insanity.—There are a few general mental states in insanity, one of which being present

gives the character and name to the disease. These are :
  a. A state of exaltation of mind, or mania.
  b. A state of depression of mind, or melancholia.
  c. A state of enfeeblement of mind, or dementia.

But one of these first two states of feeling can be present at the same time, for a person cannot at any one moment be both exalted and depressed, though he have mania to day, and afterward be so changed in his feeling as to have melancholia to-morrow, or next week, or next month.

In a general way all disease is divided into acute and chronic forms. An acute disease is one of recent origin, and from which recovery is to be hoped for ; a chronic disease is prolonged and does not tend to recovery ; an acute disease may become chronic.

Mania and melancholia are at first considered acute and curable, but, if recovery does not take place, they pass into either chronic mania or chronic melancholia, or, if the mind is much enfeebled, into a condition of dementia.

Mania.—In mania the mind is generally very active, though lacking in control, and is irregular and illogical in its action ; the patient talks rapidly, and upon many subjects, and is often incoherent, or he laughs, sings, dances, or cries, perhaps in turn ; he is often irritable and unreasonable, and perhaps threatening, and becomes more violent if interfered with.

Accompanying this mental excitement there is frequently persistent loss of sleep, constant restlessness, and great bodily activity, and indifference to or refusal of food. Sometimes the brain excitement is so great that all self-control is lost, and the patient becomes a raving maniac.

The delusions of mania are largely of grandeur and self-exaltation ; the patient thinks himself in the best of health, and very strong, or of a superior mind, or, that he is a great singer, poet, actor, or preacher ; perhaps, taking a higher flight, he thinks himself possessed of the wealth of Vanderbilt, or that he is the Pope, or the President, or even God himself.

Sometimes the excitement comes on in paroxysms, lasting a few days or weeks, with periods, more or less prolonged, of comparative mental quiet.

Melancholia.—In melancholia the expression of the face often tells the character of the disease ; the eyes are downcast, the lines of the face are lengthened, and the whole appearance is that of unhappiness.

In this form of insanity the patient may refuse to speak or interest himself in any thing, or he may moan, groan and cry, and walk back and forth wringing his hands ; when he is quiet, the mind, however, may be very active and full of delusions, which occupy it to the exclusion of every thing, driving away sleep, and making him indifferent to the taking of food or attending to his most necessary wants ; sometimes the patient talks a great deal, but always about his delusions, which are generally connected with himself, his family, or his affairs.

Melancholiacs are often tortured by fears, and, therefore, become frenzied and as wild and violent as in mania ; or they may be very suspicious, thinking that some one is persecuting them, or poisoning their food, or following to kill them. On account of their delusions they frequently refuse food, they generally sleep poorly, and are often very suicidal.

Dementia.—This form of insanity is most frequently the result of acute mania or melancholia, and comes after the force and intensity of the disease has spent itself, leaving the mind crippled and weakened. The perceptions are blunted and distorted, memory fails, the reasoning powers are weakened, the will has ceased to control, the emotions and appetites are dormant or changed, and the mind may become almost a blank, though in the narrow circle of thought there is left remains of delusions, illusions, and hallucinations. The patient is frequently careless of the ordinary necessities and decencies of life, and requires constant care.

There are degrees of dementia : it may be slight, partial, or nearly complete. During the first few months or years dementia often ends in recovery, but, as it continues, the case becomes more and more hopeless.

Monomania.—This is a term belonging to common speech, but there is not an agreement of opinion as to the existence of such a special form of insanity, nor among those who believe in it, as to what it is and what are its symptoms.

Monomania really means an insanity with but one, or, at most, a small class of delusions of the same character, the rest of the mind showing no disease. Hardly any one believes in the existence of such a narrow limit to insanity, and, getting beyond this point, there is no agreement where the limit should be set up to mark and bound it.

Some think there is a special insanity of the emotions only, and call it "emotional insanity." There is not an agreement of opinion as to what emotional insanity is ; the idea seems to be that the emotions, or one of them, so overpower reason and will as to make the person irresponsible. This condition is supposed to exist without disturbances of the intellectual faculties, and to be unaccompanied by delusions, hallucinations, or illusions. Others see in these cases no evidence of insanity ; nothing but over-indulgence of the emotions, or a want of exercise of self-control, or an excuse for crime.

Some persons believe that the appetites over-indulged become morbid and produce disease of the nervous system, and as a consequence the reason and will are weakened in relation to this indulged appetite, and the opinion is reached that it is a form of insanity. An indulged appetite for drink is called dipsomania. Others believe that unless there are present the usual symptoms, associated as they generally appear in insanity, these cases are nothing but unbridled appetites or vices.

Moral Insanity.—There are those who claim that the moral nature alone may be diseased, and the persons in whom this occurs are said to lose the appreciation of right and wrong, or have an uncontrollable propensity to do some wrong act, and take a peculiar pleasure in so doing. Special names are given to these acts, according to their character, as "kleptomania, an impulse that prompts to steal"; or "pyromania, love of setting things on fire"; or "homicidal mania, an intense desire to kill." Other persons considering these cases and finding no delusions, or intellectual disturbances, or change in feeling, thinking, or acting due to disease, call the condition one of crime only.

These are difficult matters to understand, and those who make a life-study of insanity do not fully understand them, or agree together as to what they know. They are, however, terms of common speech, and it is well to have some idea of them, as it will add interest to the study of the patients under care and charge.

CHAPTER IV.

THE DUTIES OF AN ATTENDANT.

What an Attendant Should First Learn.—The duties of an attendant upon the insane are varied, arduous, and exacting; they are associated with irritations, perplexities, and anxieties, bring grave responsibilities, and call for the exercise of tact, judgment, and self-control.

These many duties are not quickly nor easily learned, and the new attendant must be willing to fill, at first, a minor position, to begin at the beginning and learn gradually all the details of ward work; he must acquire habits of caution and watchfulness, and learn in a general way the care of the insane, before he can assume a position of authority over other attendants, the control of a ward, and the responsibility of the direct care of patients.

This last duty is the most difficult of all, because it brings the attendant into intimate relations with a class of persons, whose true appreciation of themselves, of their conditions and surroundings, is changed, whose thoughts and desires are unreasonable, whose conduct is unnatural, and who are largely controlled by insane delusions, hallucinations, and illusions.

It requires an intimate association with the insane, and a careful study of their manner of thought and conduct, to be able to successfully guide, direct, and control them.

The Relation of Attendants to Patients.—The position of attendants is often a trying one; they are liable to misrepresentation when they have faithfully done their duty; they must learn to receive with calmness a blow or an insult, or even so great an indignity as being spit upon; they must bear with provocations that come day after day, and are seemingly as malicious as they are ingenious and designing; they must watch over the suicidal with tireless vigilance, control the violent, and keep the unclean clean.

To do all this requires the exercise of self-control and kindness; the putting a curb upon the temper; the education of judgment and tact; faithfulness in the performance of duty, and a knowledge of what to do and what to avoid.

These trials are, however, but a part of the experience of an attendant in caring for the insane, for there is associated in this care much that is satisfactory and pleasurable. It is a satisfaction to know that duty has been well done; to be able to care for the sick; to do something to alleviate suffering; to tenderly watch over and soothe the dying; it is a pleasure to see a patient improving, going on to recovery, and finally able to return home cured.

Many delightful friendships are formed between attendants and patients, some lasting for years within the asylum, and some for a lifetime, with those who have recovered. Most of the insane appreciate the services rendered them, and have a feeling of gratitude for those who care for them.

Attendants should always treat patients with politeness

and respect; it is something that is never thrown away, and exerts a good influence, however rude and disrespectful a patient may behave.

Patients should not be ridiculed, their mental weakness and peculiarities made light of, nor should they be made a show to inquisitive visitors.

It is useless for attendants to try to argue patients out of a belief in their delusions, and to do so often results in fixing them more firmly in the mind. We should not however pretend to believe them, nor humor their belief, nor allow them to carry out their delusions in their dress, conduct, and general behavior.

The Character of an Attendant.—The insane should always be treated with kindness, and nowhere is the golden rule "thou shalt love thy neighbor as thyself" more necessary of application than in caring for them; and it is well for attendants, when tempted, to stop and think how, under like circumstances, they would want their mother or sister or brother treated.

Keeping this noble teaching and this high motive for right-doing ever in mind, an attendant cannot go far astray.

It is a development of character to care for the insane, and instead of being brutalizing, as some ignorant people say, it is elevating and humanizing.

Attendants should never gossip, either among or about themselves, or of their patients. It is a mean and degrading habit to indulge in; it will undermine a good character, and often become overpowering and malicious.

On the other hand, never be afraid to speak the truth, and never let a lie, or the semblance of a lie, pass your lips, or remain for a moment in your heart. Of all things be truthful.

Attendants must acquire a spirit of willing obedience, of cheerful execution of all commands and directions, and of faithful performance of every duty that devolves upon them. Unless they have this spirit, they will be unable to successfully assume positions where obedience is to be exacted from others.

They should preserve their own self-respect; in all things set a good example; be neat and tidy in their dress, gentlemanly or ladylike in their conduct; considerate of the wants and feelings of other attendants; they should "cherish a high sense of moral obligation; cultivate an humble, self-denying spirit; seek to be useful; and maintain at all hazards their purity, truthfulness, economy, faithfulness, and honesty" (Utica Asylum Rules and Regulations).

In their relation to the institution, attendants should fulfil all their engagements with the same sense of right, that they expect will be observed towards them by those who employ them. It is a business contract that is assumed, and brings with it mutual legal responsibilities, rights, and obligations. Attendants should strive to so conduct themselves, that when they leave their employment they can go away with the respect of every one, and bear with them the reputation of a good character and of work well done.

How and What to Observe in the Care of Patients.—It is important that attendants should early learn habits of close observation. The exercise of the habit increases the ability to observe, and one soon comes to see and know things he never saw, or thought of before. It is necessary to learn first the physical condition, mental symptoms, and habits of a patient, before we are able to observe and appreciate any change.

Observation, to be of value, should be systematically made, and only one thing at a time can be noticed, which must be understood before passing to another, otherwise every thing is confused.

The condition and appearance of a single part should be looked at to see what is natural, and what is evidence of disease.

In practice, written notes taken at the time, are extremely valuable in teaching close and accurate observation, and cultivating an ability to clearly express to others the result.

For the purpose of suggestion and guidance, the following system for observation is given:

Observe the effect of medicine.

The face.—Observe if it is pale, and if the pallor is sudden, temporary or permanent; if flushed, if congested, if blue with venous blood, if there are any eruptions, bruises, or scars. Observe the expression of the face.

The tongue.—Observe if it is coated, and if so, if white, brown, red, black, glazed, dry, or cracked; if it is tremulous, or drawn to one side, or protruded with difficulty.

The lips.—Observe if pale, blue, dry and cracked, if there is tremulousness about the corners of the mouth; the teeth, if covered with sordes; the gums, if bleeding.

The breath.—Observe if sweet, sour, foul, or offensive.

The respiration.—Observe if slow or fast, quiet and natural, or loud, labored, and difficult, if puffing, wheezing, shallow, or irregular.

The eyes.—Observe if congested, the color, if any blindness; the pupils, if contracted, dilated, irregular, unequal, or if they respond readily to light.

If there is cough, observe if moist or dry, if croupy, if with pain, or if prolonged.

If any expectoration, observe if it is bloody or streaked with blood, if thin and frothy, thick and purulent, or if it sticks to the cup.

The pulse.—Observe if it is slow or rapid, full, weak and thin, if irregular or intermitting. Count it.

The temperature.—Observe by the hand or thermometer.

The body.—Observe for eruptions of the skin, for sores, bruises, or deformities, or if there is any paralysis.

The appetite.—Observe if it is poor, changeable, if food is relished or disliked; if refused, if it is constantly or occasionally, and if from delusions or indifference; if there is overeating and gluttony, if food is bolted, or chewed, or if the patient has teeth to eat with.

The digestion.—Observe if natural, or painful, and if so, whether upon taking food, or if the pain is delayed; if gas is discharged from the mouth, if the stomach is sour, if the food is heavy and distressing; also observe what kinds of food give dyspepsia, and what seem to be well borne.

Of vomiting.—Observe if occasional or constant, if immediately after food, or delayed, if sour or bitter, if preceded by pain or nausea, if it contains any undigested food.

Of diarrhœa.—Observe how frequent the discharges, if with pain, and where it is situated, the color, the consistency, if there is any blood or mucus, if it alternates with constipation.

Of constipation.—Observe if alternating with diarrhœa, if habitual, the effect of medicine and food; if there are any piles.

The menses.—The quantity, if there is any pain, its cessation and reappearance, if any effect upon the mental condition.

Of pain.—Observe the character and severity, its location, and any evidence of a cause.

Of dropsy.—Observe if it is general or local, if in the chest, face, abdomen, arms, or legs; if there are any varicose veins.

Of Sleep.—Observe the length of time, if quiet and natural, if restless, if deep or light, if there is great drowsiness or continued wakefulness, and the effects of medicine.

Of unconsciousness.—Observe if it comes on slowly or suddenly, if partial or complete, if the patient can be aroused.

Of convulsions.—Observe if slight or severe, if of short or long duration, if continued or interrupted, if general or of one side, or of an arm or a leg, or the face, or of a few muscles only.

Of the mental condition.—Observe if fixed or changeable, the nature of delusions, illusions, or hallucinations; dangerous attempts or threats toward himself or others; any change in the mental state.

Of habits.—Observe if fixed or changeable, how formed or how corrected.

Of the general conduct.—Observe the dress, if neat

and tidy, or otherwise, private habits, care of personal wants, improvement in conduct, the influence of attendants and other patients, or the influence the patient himself exerts on others.

This by no means includes all that it is necessary to observe, but it contains much that is important, and the system, if studied and used practically, will suggest to the observer whatever may require attention.

The Control and Influence of Attendants over Patients.— By a "smart attendant" is meant one who sees little to do beyond having a control of the ward by a rule that is close and exacting, who maintains a strict discipline, and who has a love for cleanliness, order, work, and scrubbing. But a "useful attendant" is one who tempers these mentioned traits, by striving to gain the confidence of his patients, by exerting over them a beneficial influence, who is able to bring the individual patient into accord with his surroundings in the asylum, so as to help his improvement or recovery, meet his wants, and increase his comfort and enjoyment. In order to do this it is necessary that the attendant should give careful study and attention to each patient. Such a study will soon demonstrate to, and teach the attendant the fact, that the insane are very individual in their habits, and while no two are alike, there are resemblances that in an asylum are made the basis of classification by wards: there is the convalescent, the suicidal, the demented, the sick and feeble, and the noisy or violent wards.

Attendants must first learn that patients are not to be treated merely as a ward full of people to be kept in order, to be clothed, fed, and put to bed, but that the peculiarities of each patient are to be studied, and that it is their duty to know thoroughly the wants, and condition of each case, and how best to care for and control it. The better knowledge an attendant has of the individual, the better he can care for a ward full of individuals.

The persons who are under our care are always to be considered as patients, and it must be remembered that these sick people are sent away from their homes and given over to us, though strangers, because it is supposed that we can do better by them than their friends are able to do.

Their position is one of helplessness and dependence upon those who are placed in charge, and we are properly held responsible by the friends and the public, for a judicious exercise of the power and influence we possess over them.

Patients are not rightly influenced by the mere exercise of authority or by dictation or command ; these they fear and obey, or resent and resist ; but we should always appeal to the highest motives for obedience and correct conduct, and we should lead our patients to trust and not to fear us. In our dealings with them we should be truthful, straightforward, and strictly upright, and exercise over ourselves patience and self-control.

We can generally control our patients by the exercise of sympathy, kindness, and tact, joined with a reason for what is required, and where more is needed, a firm, kind authority and command will suffice.

The use of authority, restriction, and restraint is to be avoided, while on the other hand patients are to be allowed all the liberty and freedom they can safely enjoy, and taught to exercise all the self-control they are capable of.

The granting of more freedom and liberty of action than was formerly accorded the insane, does not imply a change in the character of the disease, but improved methods of care, and places more responsibility upon the attendants. The degree of liberty to be allowed must, in each case, be decided by the physician, and the attendants should closely observe the patient, and report any symptoms which makes the enlarged freedom dangerous to the patient or to others.

Patients being sick, are sent to the asylum that they may be kindly and judiciously cared for, and, if possible, cured. As many patients who may never fully recover may so improve as to be able to return to their homes, and, as it is impossible to say that any given patient will not recover, each case deserves and should receive our best care and efforts to this end.

Because our patients are sick they must be nursed, and nursing means tender care. And it is a nurse's duty to do all in his power to alleviate pain and promote bodily comfort. The insane are subject to all the ills that flesh is heir to, and there is always among our patients much sickness and bodily suffering. Many patients cannot tell when they are sick, nor when they suffer pain, but they show sickness and pain, and often appeal by their manner for that care and sympathy, we all feel in need of at such times.

These silent symptoms should be observed by the attendants, who should always see and know when their patients are sick. Some of these symptoms are, crying, moaning, weakness, going to bed, or lying down, cough, changes in respiration, signs of fever, a flushed face, quick pulse, or chills, a pale face, vomiting, or diarrhœa, and loss of appetite.

Much insanity is associated with great physical disturbances which require careful nursing. The old and feeble, the paralytic and bedridden also require special attention and care.

From this it appears that the care of the insane calls for the exercise of self-control, habits of close observation, the using of good judgment, the putting forth of ennobling influences, and the tender care of the nurse.

CHAPTER V.

THE GENERAL CARE OF THE INSANE.

THE Reception of New Patients.—Attendants must at once study the peculiarities, the physical condition, and the mental symptoms of a new patient, so as to know the case thoroughly.

New patients should receive special attention ; their fears quieted ; they should, if in a proper condition, be introduced to the other patients ; the effect of being in so large and strange a place, where the doors are locked and the windows guarded should be noticed, and unpleasant impressions overcome ; they must be told they have come among friends and will be kindly treated.

The necessary rules of the ward should be explained ; they should be invited to their meals, shown to their rooms and told at bedtime the night watch will visit them, and they must be assured that no harm will come to them.

The first impressions a new patient receives may be the lasting ones, and influence their whole conduct in the asylum. If they resist what is necessary to do for them, do not struggle and contend with them, and force them to bed, or to the bath, but first seek advice from the supervisor, or the physician.

Always search new patients, unless otherwise ordered,

for money, jewelry, weapons, medicine, and other like articles, or if in doubt what to do ask for directions. The head, body, and clothing should be examined for vermin, and the body for injuries and bruises. If what is wished to be done in this particular is explained, patients will generally quietly allow it.

Work, Employment, and Occupation.—By this is meant whatever occupies the patient's time and mind, in useful and pleasant ways.

Of all things idleness and loafing are the worst; even games, such as billiards and cards, if indulged in to the exclusion of useful employment, will degenerate a patient.

Some willing patients are kept in a tread-mill of daily work, their monotonous life never broken by a diversion, an enjoyment, or a hope. It is very questionable if it is beneficial to make a patient drudge through such a daily routine.

Asylum life should be made as home-like, pleasant, and natural as possible; as a rule every patient who is able should do some useful work every day, and to this should be added the diversion, that comes from amusements and the enjoyment of innocent pleasures.

Occupation then means a great deal more than work; it is the way a patient spends his time. Unless encouraged and directed, patients may occupy themselves in thinking of their delusions, in noise, violence, or destructiveness, in idly walking up and down the wards, in the indulgence of secret vices, in gossip, in spreading discontent, in prayer, or in constant Bible reading. Some patients really work hard trying to do nothing, and have no more ambition than to sit around on the ward, and chew tobacco, and indulge in idleness.

Patients should be encouraged to do something for themselves, the women to make and mend their own clothes, to keep their rooms in good order, and assist about the ward. They should be made to feel that they can add to their own comfortable surroundings by their own efforts.

For the men, ward work is not so natural or tasteful, but they will do with interest much of this kind of work; to this may be added employment in decorating their own rooms or the ward, and in caring for plants and flowers.

The women can add to ward work, sewing, knitting, mending, embroidery, artificial flower making, quilting, care of flowers in the ward, and it is often a real enjoyment for patients to make some little present for their outside friends. The laundry offers an inviting field for some patients, but it is often too hard work, especially when they are sent twice a day to the wash-tub, or kept in the hot ironing room. A half day is enough for most patients, and many are not strong enough to go there.

Out-of-door work is well suited for the men. The farm, garden, lawn, barns, and machine-shops offer much that can be made useful for the patients' employment; the different mechanics and artisans about the asylum should have patients working with them.

Thus it appears there are many directions for patients to work, and it is also true that all patients are not suited to do the same work nor the same amount of work. Whatever they do should be for their benefit alone, otherwise we might take a contract for a given number of patients to work a given number of hours every day, a good deal as has been done in prisons and reformatories, but no one would believe such a course for the interest, improvement, or recovery of the patients.

The only rule to go by is, that the work and occupation shall be for their own good, and, that they shall not be made or encouraged to work for any other purpose.

As a rule, patients should be allowed to employ themselves in ways that most interest them, provided it is useful and seems to be beneficial.

Over-work is as bad as idleness; too much sewing will often give a sleepless night.

Generally all patients may be allowed to engage in light work, without special directions; new patients, however, should not be sent off the ward, or given tools that may become weapons, unless by order of a physician.

It is a bad habit for attendants to sit idly by, or stand around with their hands in their pockets, and have patients do all the work. It may be so necessary to watch the patients that the attendant cannot work steadily, but he should have the appearance of doing something, and if possible join with them in work.

A party of women sewing, should be laughing, talking, telling stories, perhaps singing; they should be made to enjoy the time, and not to look upon it as something irksome.

Some patients are too feeble in mind, and some too feeble in body to work; many need rest, quiet, and nursing, and directions for the care and occupation of such patients should come from the physician.

Many of these patients will do a little, others can be amused, or read to, and their minds thus diverted from their troubles, and turned into pleasant and cheerful directions of thought.

It has been shown that work is not the only useful way that patients may occupy their time, that nothing but work is as bad as no work, and that they should have diversion, enjoyment, and entertainment.

For the entertainment and occupation of patients, there are furnished, dances, concerts, theatricals, billiards, cards, pianos, books and papers, schools, chapel services, walks, rides, and excursions, and they also receive visits from friends, and write and receive letters.

Patients should be encouraged and sometimes made to take part in these natural and pleasant amusements; of course every patient cannot play the piano, or billiards, but among these many forms of recreation, all patients can find ways of diversion and means of enjoyment.

Thus early in the study of the care of the insane, it is learned that the life of patients is to be stripped, as much as possible, of restriction and restraint; that self-control is to be taught; that useful work is to be encouraged; that amusements and innocent pleasures are to be enjoyed; in a word, attendants are to learn, that the characteristics of institutional life are to be lessened, and those of a home life made prominent.

The Patients' Care of Themselves.—The general tendency of the insane is to mental enfeeblement, to neglect of person, and to slovenly habits. Patients should be encouraged as much as possible to care for themselves; to be helpful towards others; to do such work as they are able ; to seek amusements, and to live as much as possible such a life as we ordinarily are accustomed to outside the asylum.

Patients should be encouraged to keep themselves tidy, and nicely dressed, to have the care of their clothing ; if possible, they should be given a room of their own, which they should take a pride in keeping in order, and ornamenting with pictures and flowers ; and should be allowed to do whatever will help maintain their self-respect, self-care, and a feeling of individuality.

There is great difference in patients as shown in their capacity for self-help. Some seem to be able to do nothing, some everything. Nothing can lighten the burdens of attendants so much as to make the helpless self-helpful. Nothing benefits the patients more. Do not abandon effort for any patient. Unexpected and gratifying results are the rewards of earnest efforts.

Out of Door Exercise—Walking.—If possible, patients should be out of doors every day. In the summer much time can be spent in the fields, on the lawn, either walking or sitting under the trees ; in the winter time shorter walks only can be taken, but on pleasant days, often an hour may be spent out of doors. Warm clothing and good shoes must never be neglected, and the person must be thoroughly protected, because the insane are frequently " cold-blooded," that is, the circulation is poor, the hands and feet congested, blue, and cold, they make animal warmth slowly and with difficulty, and easily suffer from the cold.

Many patients go out to walk on parole. Those who are allowed this liberty will be designated by the physicians; any change in the patient that makes such liberty dangerous should at once be reported. Others go out in large parties, with few attendants to care for them, while the old, sick, and feeble, the homicidal and suicidal, the noisy and violent, require special care and attention in their exercise and walks.

Clothing of Patients.—In many asylums each patient has his own clothing. Every article should be plainly marked with his own name, and should be used only by the patient to whom it belongs, and never given to any one else to wear. All clothing should be kept clean and well mended, and should be properly put on and kept on during the day. There should always be enough to keep the patient warm, and changed with the changes in the weather, or the temperature of the ward, or the needs of the patient. The sick, feeble, and old always need extra clothing; that worn next the skin should be changed at least once a week, and all clothing should be changed as often as soiled.

Bathing of Patients.—Every patient should be bathed once a week and as much oftener as is necessary. The tub should be cleaned and the water changed for each patient; the temperature should be about ninety-five degrees, or not hot to the hand, and the tub should be about two-thirds full. The head, neck, and body should be washed with soap; each patient should have a clean towel, be wiped dry, and given a change of clean clothing.

Some patients object to bathing; they fear the tub, but will wash with water and a sponge, and they should be allowed to do so. Others want to bathe first; let them if possible. Others will not bathe the day the rest do; it is sometimes best to humor them.

Some patients have to be forcibly bathed. In such cases always wait, use every art to induce them to bathe, and before acting send for advice.

Attendants are too prone to think that every thing should be done by rule, and that all must be forced to obey the rule. Most will observe it without trouble, and the object sought can generally be gained by patience, tact, and kindness.

Serving of Food.—The dining-tables should be neatly set and made attractive ; the food should be promptly served, and while hot ; all patients should be at meals, unless excused by the physicians. Economy should be practised, and every thing should be used or saved. Each person should have enough, but no one should be allowed to make a meal of a delicacy, or take all of the best of a dish. Some patients would waste a pound of butter or sugar at each meal ; enough is sufficient for anybody.

The old and feeble should be served by attendants ; those without teeth should have their food prepared, and the meat should be cut very fine. Those who will not eat must be kept in the dining-room and fed ; the attendants may use force by holding the hands, and placing food in or to the mouth ; but it is dangerous to do more, and holding the nose is something that is never allowable. If these efforts to get them to take food do not succeed, report to the physician. Some patients from delusions will eat certain kinds of food, and either not get enough or not a sufficient variety.

A mixed diet is the best, and patients should if possible be made to eat bread, butter, meat, vegetables, and drink milk and plenty of water. No patient should be allowed to lose in flesh and strength on account of failure to take sufficient, or proper food ; before these things happen it should be reported to the physician. Some patients will only eat enough if they are allowed to eat it in their own way ; they will eat it perhaps standing, or after the others have finished, or alone, or in their room, or they may steal it, if given the opportunity. Such peculiarities often have to be indulged.

Some patients will take nothing but milk, then about three quarts a day are needed ; eggs may be added and are often readily taken, and some may be got to eat bread and milk, which is a very nutritious diet.

The food of the sick should be nicely and invitingly served, and efforts should be made to meet their whims and fancies.

Patients who are so profane, violent, or noisy, that they are not allowed to come to the dining-room, must always be fed by, and in the presence of an attendant, and meals should not be passed into a patient's room and left there.

Knives and forks should always be counted by an attendant before and after each meal ; care should be used that they are not lost, secreted, or carried out of the dining-room by patients. No one but an attendant should ever handle the carving knife and fork, or the bread knife.

Care of Patients when Going to Bed, or Rising.—The beds should be daily aired, and always clean and nicely made up ; for a filthy patient a straw bed, that can be changed, alone is clean.

All patients do not need to go to bed at the same time, and while some are able to care for themselves, most need care, attention, and watching. The helpless should be dressed and undressed, and put to bed first: the violent and homicidal need to be watched, and should be put to bed early, while the suicidal should be kept under supervision, and put to bed at the most convenient time. After a patient is in bed, an attendant should go into the room, with a lantern, so as to see that every thing is in order and safe, and, with a cheerful "good-night" close the door. Patients who need care should be visited during the evening, and left clean and in good condition to be cared for by the night watch.

In the morning patients need attention before any thing else is done. First, the suicidal, sick and feeble, the violent, and those likely to be filthy should be visited, and every patient should be washed and dressed before breakfast; or, if for any reason they do not come to this meal, their faces and hands should be washed, the bed put in order, and the room made clean and aired.

After these things have been attended to, the ward work should be done, though generally the two can go on together.

Care of Patients during the Night.—After the patients have gone to bed the ward should be quiet, doors should be quietly closed, voices lowered, and loud calls and laughter not indulged in, squeaking boots should not be worn, and heavy walking avoided. Many patients go to sleep early, but are easily awakened, and may remain sleepless till morning, or at least a part of the night.

The night watchers have responsible, arduous, and trying duties. Attendants should always, during the night, quickly respond whenever a demand is made upon them for assistance, though an unnecessary call should never be made. The night watchers should be informed of any changes that have occurred during the day, that will require their attention during the night; they should see new patients and be made acquainted with their peculiarities; they should visit the wards during the evening before they come to the medical office to receive instructions from the physicians.

It is the duty of a night watch to visit regularly all the wards under his charge; to see and know the condition of the sick, the helpless, feeble, the suicidal, and the epileptic; to attend to, by taking up, those who are inclined to be filthy, and wash those who need it, and make them, their beds, and rooms perfectly clean. He should observe the conduct of new patients, be watchful of the violent, know how much wakeful patients sleep, visit all associated dormitories, wait upon all those who need attention, and guard against fire and accident. The night watch should place each day on the medical office table, a detailed account of every patient that needed care or attention, who was disturbed, or did not sleep during the previous night.

Patients should be left clean for the night watch, who should leave them in as good condition in the morning, for the day attendants, and any neglect in these directions should be reported by either party. Sick patients frequently have to receive special night service, to be watched, and given food and medicine. When this cannot be done by the night watch, it devolves upon the day attendants, and is a duty that should be cheerfully rendered.

During the night, any accident, attempt at suicide or to escape, or unusual violence, persistent sleeplessness, or being out of bed, a serious sickness or change for the worse, or the approach of death, should be reported to the physician. It is, in many institutions, the duty of the night watch to report any neglect or misconduct on the part of an attendant or employé, and it is something that should be faithfully and impartially done.

Having briefly sketched the general duties of an attendant, it seems best to again remind them, that an asylum is built and maintained for no other purpose than for caring for the insane; that each patient is entitled to the best our means can afford; that while the attendants are not responsible for the medical treatment, they are for that kind and intelligent care it is within their province to give; and they are also reminded that, so far as it can be done, such personal attention is to be given to each patient as will assist in recovery or improvement, or promote his well-being.

CHAPTER VI.

THE CARE OF THE VIOLENT INSANE.

A CAREFUL study of each violent patient, of his habits, delusions, and hallucinations, of his peculiar manner of showing violence, and a knowledge of what is likely to provoke outbursts is necessary to properly care for him. An attendant's ability to successfully manage a ward full of patients will depend largely upon the study given to, and the thorough understanding of, each case. Such study will soon teach him that every violent patient has peculiar and pretty constant ways of showing and exercising violence, and that the same rule of individuality holds good among this, as it does among other classes of the insane.

Having learned what will cause violence, it can often be avoided by removing the cause; having learned the symptoms that precede a patient's outbursts of violence, they can sometimes be averted, or preparations made to control them; having learned in what direction violence is shown, how sudden, blind, or furious it may be, or how slow, deliberate, and planned, the attendant is better able to meet, manage, and control it.

Few patients are so continuously and furiously violent as to need constant repression, and the directions how to care for such patients can always be given by the physician. Most violent patients are subject to the firm, kind control of attendants, and can be kept sufficiently quiet and orderly; they should never be left alone, and mops, pails, brooms, chambers, and all other articles, that may become weapons should not be left within reach. Strong comfortable clothing can generally be kept on the most violent and destructive, with care and attention from attendants, but not without.

Many violent patients will employ themselves and be the quieter for so doing. Light out-of-door work is the best employment for this class, and out-of-door walking and exercise should never be neglected. On the woman's ward knitting, sewing, mending, and ward work are suitable for many, while some will work at the laundry, and others will go quietly to church and entertainment; books and illustrated papers should be furnished and will be much read and enjoyed.

As a rule the more violent patients are restricted, kept continuously on the ward, or in a small room, and given no work, amusements, walks, and exercise, the more noisy and violent do they become.

Attendants must learn that mere noise, and much of maniacal activity, such as running about, jumping, or pounding, is not in itself harmful, and that unless such patients are doing themselves injury, or so disturbing the ward and other patients as to require interference, it is better to control than to repress and restrict them.

Many violent patients are subject to such paroxysms of great violence as to require immediate care and often temporary control at the hands of attendants. Generally these paroxysms spend themselves after a short time, but if they do not, advice and help can be called for.

By careful watching, the approach of these paroxysms can be known and often avoided. This may be done by removing the cause, which is often the irritation of another patient or an attendant, by a word, a joke, by simply letting the patient alone, or by a firm show of authority, or by any other means experience has taught to be useful in the particular case.

If necessary to hold a patient, three persons should be able to care for the most violent. This can be done by grasping each arm at the wrist and elbow, and holding it out straight, the attendants standing behind while another passes the arm about the neck and holds the chin, to prevent biting and spitting; the patient may then be walked backward and seated in a chair

After the violence has subsided, though the patient should continue to scold, swear, threaten, or cry, he should, as soon as possible, be left alone, the attendants walking away, but remaining watchful. Do not, unless it is necessary, interfere to stop the noise, for it is often a substitute for the violence, and the attack wears itself out in this way.

If necessary to carry a violent patient, it can be done by four or six attendants. The face should be turned downward, thereby lessening the power to resist, and, to prevent dislocating the arms, the patient should be carried by the shoulders and chest; the bands about the neck should be loosened.

In using force in the care of violent patients, it should always be done as gently as possible, and struggling should be avoided ; he should never be choked or kicked, receive a blow, or be knocked down ; the arms should never be twisted, nor a towel held over the mouth, but if the patient persists in spitting it may be held in front of the face.

Care must always be used not to injure a patient while exercising necessary control. In the violence of a patient innocent injuries are sometimes received. The attendant is excusable if he can show that he used necessary force only, without malice.

A violent patient should never be struggled with alone, and on a well-managed ward help will always be within call. It may be necessary, however, to break this rule in order to prevent homicide or suicide, or serious injury to another patient, or setting the house on fire.

It is better not to visit the room of a violent patient alone, and if an attack is feared, especially with a weapon, the door should be slowly opened, and held so it can be quickly closed. The patient usually makes an immediate attack, and, before he has recovered for a second, can generally be disarmed and controlled.

Violence usually consists of noise, tearing the clothing, breaking glass or furniture, biting, scratching, striking, hair pulling, kicking, or attacking others with weapons. It is sometimes secretly and deliberately planned and skilfully executed, though generally without reasoning or direction, but blind and fierce.

The care of the violent insane involves the careful study of each case, with constant watchfulness, and the exercise of a control that is kind, but firm and unyielding, that does not repress except when necessary, nor restrict without reason, that indulges whenever possible, that never drives, scolds, or threatens, but influences, guides, and directs. The greatest liberty possible should be allowed, and self-control encouraged, and work, occupation, and amusement should be furnished. An attendant must always remember that fear is the lowest motive to govern by, and that kindness will often be appreciated and returned.

Care of the Destructive Patients.—Besides the violently destructive patients, there are some who are maliciously destructive, and who exercise all their ingenuity to escape the watchfulness of the attendants; who glory in their wrong-doing; who openly say they cannot be punished, and exultantly tell the physician how they have outwitted the attendant, or proclaim before him his shortcomings and neglect. Such patients will destroy their own or others clothing, they will steal and hide, or throw it out the window or down the water-closet, or erase the name by which it is marked. They will destroy bedding, windows, crockery, pictures, or furniture. With a pin, a nail, or a bit of glass or wood, they will mar and deface their room or the ward, and often do damage that cannot be repaired. The only way to meet such cases is by watchfulness. They should be kept, if possible, at work, or at least with a company of workers, and therefore under constant observation. When put to bed their clothing, mouth, hair, and person should be thoroughly searched. Kindness often has but little effect, but a threat is apt to make them more determined to destroy.

The Care of Patients by Mechanical Restraint and Seclusion.—All the restriction of an asylum is restraint. The locking of bedroom doors at night is very restricted restraint. Most patients in an asylum have a feeling that they are under great compulsion and restraint, in being deprived of their liberty. It has already been taught that patients are to be given all the liberty possible, that restraint over their freedom is to be exercised no more than is absolutely necessary, and that the greatest good of the patients alone is to be thought of.

These teachings are equally true of special forms of restraint. If used at all they are to be used for the good of the patient alone, and an attendant should be able to care for any case without restraint.

Restraining apparatus should never be kept on the ward. An attendant should never ask that it be used, nor say he cannot get along without it.

If ordered by the physician it is the attendant's duty to see that it is so applied as to do no injury, that it does not bind or tie the patient down, that it does not irritate and make the skin sore, nor restrict the free movement of the limbs.

Patients who are restrained are not to be further confined to a chair without specific order. Restraint used during the day is not, unless so ordered, to be continued at night nor reapplied the next day. Patients are to be taken frequently to the closet. Restraint should be taken off several times a day, and kept off long enough to give relief to any feeling of discomfort, and free movement should be allowed. When patients are restrained they need unusual care and watching, and should never be left alone.

The attendant should be informed why restraint is used, and what is hoped to be gained by its use. He should closely observe the effect upon the patient and compare his condition with what it is when not restrained. The result of these observations should be reported.

Thus used, an attendant will soon learn that it is not the easiest way to care for a patient, that its use involves increased watchfulness and care, and greater discretion, and that it is strictly a form of medical treatment. It is a harsh remedy at its best, and needs to be used with kindness, intelligence, and judgment, and it is to be applied but for one purpose, namely, that the patient may be benefited.

The Use of the Covered Bed.—Like restraint it is never to be used except by the orders of a physician, nor is its use to be repeated without special orders; it is always to be considered a method of treatment and something the attendant has no interest in, except to know how best to use it when ordered to do so.

When in a covered bed the patient should be frequently visited; he should be taken up at least once in three hours, unless asleep; the bed and the patient should be kept perfectly clean. If used in the daytime an attendant should sit beside the patient for some hours and try to keep him quietly in bed, and the same should be done in the evening when the patient is put to bed. An attendant should be able to report how much the patient sleeps, how much quiet and rest is obtained, the effect of the treatment, and compare the condition of the patient when in the bed with what it is when not used.

The Use of Seclusion.—Seclusion is shutting a patient alone in a room in the daytime. If allowed to be done without orders from the physician it should be immediately reported. If ordered to be continued the patient should be seen at least once in fifteen minutes, while many need to be seen once in five minutes, and an attendant should never be far from the door. The patient should be frequently taken to the closet. The effect and result of seclusion should be observed and reported.

Many physicians never use any form of restraint, while others make considerable use of it as a means of treatment. An attendant should be able to successfully care for any case, so as to meet the wishes and directions of the physician, and only as he is able to do this can he give the patient the highest standard of attention, care, and nursing.

CHAPTER VII.

THE CARE OF THE HOMICIDAL, SUICIDAL, AND THOSE INCLINED TO ACTS OF VIOLENCE.

PATIENTS with Delusions of Suspicion demand special care, and are properly classed with those inclined to commit acts of violence, because they are frequently fully under the control of delusions, which make them dangerous and difficult to manage.

Many patients have ideas that make them suspicious of those about them ; these may relate to the patients, but more frequently to the attendants and physicians, and they may arise from delusions, hallucinations or illusions. This class of patients is apt to be morose, cross, and irritable ; they sit brooding over their fancied wrongs ; repulse advances and friendly intercourse ; they refuse to employ themselves, and do not respond willingly to the requirements of the attendants.

Our most trifling and unmeaning acts may give rise to the most intense suspicions and hatred. A look, a shrug of the shoulder, the manner of shaking the head, a cough, the squeaking of our boots, are frequently enough to arouse these feelings.

Suspicious patients often think they are the subjects of ridicule ; that their thoughts are read and proclaimed to the ward ; that their virtue, truth, or honor is called in question, and the accusations openly told to others, or that they are called vile and insulting names. They often have delusions of conspiracy to do them or their families harm, and connect the attendants and physicians with them, thinking, as they keep them locked in the asylum, they are associated in the conspiracy. Sometimes these patients think themselves some great persons, perhaps that they are a member of the Deity, or a ruler, or prophet, or that they have some great mission to perform, and that they are deprived of their rights, or their work interfered with, by being kept in the asylum, and that those in authority are imprisoning and persecuting them. Such persons may be, on account of their fancied wrongs, very suspicious, and even violent towards those who care for them.

Other patients have suspicions and fears of bodily harm. They may think they are to be tortured, that they are to be burned alive, or that some one is trying to kill them. To-day, as I wrote these lines, a patient told me she did not sleep last night for fear the night-watch would kill her—saying that God told her the watch was armed with a knife for that purpose, and she threatened homicidal violence in defending herself.

Many patients mistake ordinary sensations of pain and bodily discomfort, and have delusions that they are being injured. The feelings of dyspepsia may make patients think they have been poisoned; ordinary pains or aches, that they have been shot, stabbed, or pounded; women may, for some such causes, think they have been violated or are pregnant. Peculiar sensations of various kinds may make patients think some one is affecting them by electricity or mesmerizing them.

It is very easy to trace from such ideas of persecution and suspicion, the origin of homicidal, suicidal, incendiary and other violent tendencies and acts.

Homicidal Patients.—Patients are sometimes both homicidal and suicidal, and sometimes they are inclined to only one of these forms of violence. Homicides are not of frequent occurrence in an asylum. The better the care the less is the liability to homicide. But there are always a great many homicidal patients, and many more who have delusions and ideas that may cause such tendencies to arise.

Many patients are homicidal merely from violence and frenzy, and without any settled plan, any fixed delusion, or intense suspicion. They may attack others suddenly and furiously ; they may commit the act while trying to escape, or it may be the result of the violence of acute mania. Other patients become homicidal under the desire to protect themselves from supposed assaults. They may think a person who is approaching them is coming to kill or torture them. Others are homicidal from any of the ideas of persecution and suspicion that have just been spoken of. Sometimes patients hear voices telling them to commit the act, perhaps it is God's voice commanding a father to offer up his only son as a sacrifice, or a mother to kill her little children to save their souls, or keep them from some misery or crime that awaits them. Patients may think themselves God, or a king, or ruler, and therefore have a right to take life. Homicidal patients are often among the quietest, and are found in the quiet wards. They frequently lay careful plans, are secretive, and only try to commit the act when they feel sure it will succeed.

Patients who are homicidal should be especially watched. They should, if possible, be kept employed, but never given tools that may become weapons. They should sleep in a room by themselves. All persons against whom they have delusions should be warned. Patients against whom they harbor suspicious or homicidal ideas should be separated from them.

Attendants should remember that a mop, a pail, or a chair, may become a dangerous weapon, or that a knife, scissors, or a sharpened piece of iron or tin, may make a fatal wound.

Suicidal Patients.—Patients with this tendency will generally talk freely of their suicidal ideas, tell why they wish to commit it, what provokes the idea, and how they would do the act. They are frequently grateful for the care bestowed to help them resist the impulse, and will sometimes tell the attendants when they feel the suicidal ideas coming on, that they may be the more surely watched.

Melancholic patients are most inclined to suicide, but any insane person, whatever the mental state, may commit the act. Delusions of depression generally cause the suicidal ideas, but hallucinations sometimes play an important part. Some persons are simply tired of life, and see no hope in living; some think they are a burden to their friends, and that they are taking food away from their children; others wish to die to escape from their misery, which is generally a mental, and not a physical suffering; others that by so doing they may get forgiveness of their sins; others because they think they will save their children from a fate like theirs; sometimes it is the result of hallucination, as a direct command from God, telling them to commit the act.

But few patients are constantly determined to commit suicide. The opportunity offered, as a bath-room door left open, a rope, a knife, often prompts the desire and allows the accomplishment of the deed.

Attendants must remember that it takes but a few minutes to commit suicide, by drowning or hanging—but a moment to cut the throat; that persons can drown themselves in a pail of water, hang themselves by the hem of the sheets, cut their throat with a piece of glass or tin. Sometimes patients slyly save their medicine until they get enough to poison themselves.

About dusk in the evening, or at early morning, is the time when patients are most inclined to commit suicide. When patients are rising, going to bed, or to their meals, when going to chapel, amusements, or to walk, when all is busy and astir on the ward, are the times that offer the most favorable opportunities for the act.

Often patients have a certain way by which they will commit suicide, and they will do it in no other; one wishes to drown himself, another to hang, and another to take poison. Sometimes patients will appear cheerful to avoid suspicion and so find their opportunity, while others may suddenly and while convalescent commit the act.

The only way to care for patients who are suicidal, is by constant watchfulness day and night. During the day they should be employed and kept with other patients, they should be especially looked after at those times when opportunities for suicide are increased. At night it is better to have them sleep in an associated dormitory with some one to watch them. If a patient is found hanging he should at once be cut down, all restriction about the neck removed and artificial respiration set up, or if drowning, the mouth and lungs should be first emptied of water; if there is hemorrhage compression should be made upon the artery, or if this is not possible, then directly upon the wound. How to control hemorrhage and do artificial respiration will be described in the chapter on emergencies.

Patients Who Have Tendencies to Self-Mutilation.— Some patients horribly mutilate themselves. They may gouge out an eye, cut off a hand, pull out their tongue, or even disembowel or dreadfully burn themselves. Some patients persistently beat their heads against the wall or floor, others scratch the skin, making large sores. Such patients frequently think certain passages from the Scriptures apply to them, and they must obey the application and command. They quote in justification of the acts, "An eye for an eye," "And if thy right eye offend thee, pluck it out," "And if thy right hand offend thee, cut it off." Talk of this kind should make an attendant very careful and watchful of the patient.

The origin of the ideas that lead to the attempts at self-mutilation is to be found in delusions, and arise in the same way as do ideas of suicide and homicide. These patients are all of the same class and need the same character of care, attention, and watching.

Patients with Tendencies to Setting Things on Fire.—
Patients with these tendencies generally desire to commit incendiary acts under the influence of delusions or hallucinations ; added to these there are frequently suspicions and feelings of wrong treatment, and the patient takes this way of showing revenge, or, as he may say, of repaying the wrong. Sometimes patients are so feeble in mind that they light a fire because they think it is a pretty sight to see it burn. There are some conditions accompanying epilepsy where patients are liable to commit any of the class of violent acts described in this chapter. The special care demanded by these patients will be fully spoken of hereafter.

There are some patients whose minds are so distorted by disease that they seem to take a pleasure in wrongdoing, and are much inclined to do great mischief, and sometimes to commit acts against life or property.

The care demanded by patients who are inclined to acts of violence is practically the same for all. The attendant should thoroughly know the habits, peculiarities, and delusions of each person under his care; he should exercise constant watchfulness, and remember that a moment of thoughtless inattention may give the opportunity for a patient to commit some violent act, that will cause him lasting regret. The mind of a faithful attendant will, when upon duty, always be full of anxiety, and there should be in the care of very troublesome patients of this class frequent relief.

CHAPTER VIII.

THE CARE OF SOME COMMON MENTAL STATES, AND ACCOMPANYING BODILY DISORDERS.

CARE of Patients in the Earlier Stages of Insanity.—Patients in the earlier stages of insanity act differently, one from the other, when first brought to the asylum and placed under care and restriction. Sometimes patients accept the situation and fit into asylum life without any friction. They may even come willingly, knowing they need care and treatment, or from confidence in their friends or their physician's advice.

To some patients the restrictions of an asylum are irksome and misunderstood; the quiet, regularity, and routine of the life on the ward does not at first affect them; they may, and often do, become fretful, are irritated by their confinement, sleep poorly, eat little, and may make violent efforts to escape.

These conditions, if nothing is done to occupy the patient's time and mind, and so relieve them, will often be sufficient to provoke violence. These patients should be carefully watched and their condition studied; they should be brought under the kind control and influence of attendants, induced to take part in the regular order of the ward, and, if strong enough, should be furnished with proper work and occupation.

Patients, when first brought to the asylum, frequently have much anxiety about their homes, their families, or their business affairs. This is particularly true in recent cases of insanity, because such patients often have cares and responsibilities, or they have tried to continue to assume them, up to the time of coming to the asylum. Special care should be taken to quiet fears in these directions; they should be assured that they are groundless, told they will be allowed to communicate with their friends, that they will be visited by their family, and that all their interests will be cared for.

It is impossible to speak of the varied causes of insanity, or of the equally varied manifestations of the disease and conduct of the patient at its onset, but there are a few conditions which, being present, give a character to a particular case, and suggest the care required.

Sometimes, as has been said, the patient partly realizes his condition, and is willing to come to the asylum, and in every way to conduct himself in accordance with the rules and requirements.

Sometimes the onset is slow and the symptoms so obscure as to attract little atttention. Following this, more decided symptoms may appear; the patient may become violent, noisy, destructive, or sleepless, or he may try to commit suicide or homicide, or do some other act of violence; or the great restlessness, moaning, crying, and sleeplessness of melancholia may come on, or the patient may refuse, for several days, all food. The reason for bringing such patients to the asylum is that they can no longer be kept at home.

Following the treatment that has been described, these

patients will frequently in a short time become more quiet, self-controlled, and more easily influenced and cared for.

The earlier stages of insanity are frequently accompanied by considerable disturbance of bodily health. The appetite is poor, the digestion disordered, the bowels constipated, the breath foul, the secretions of the skin changed and often offensive, the temperature a little elevated, the pulse rapid, and the heart weak. Sometimes, on the other hand, the temperature is normal, or a little below, while the hands are cold and clammy. In addition, nutrition is frequently impaired, so that the food taken by patients does not seem to properly nourish and strengthen. All of these symptoms are not present in a given case; sometimes most of them may be, and again but few are to be noticed.

The important lesson to learn in the care of these cases is that such patients may rapidly pass into a more serious condition, in which there is great exhaustion, which is always alarming, and may even result fatally.

Recent cases, such as have been spoken of, need our best care, closest attention, and kindest nursing. The patient should daily take sufficient food, which, if necessary, should be enforced, and the opportunity for sleep promoted. A few days, or a day, without food and sleep may bring on alarming symptoms.

For these patients, milk is the best article of diet; it is most easily given and readily taken; it should be given by the glassful, or if not able to do this by the spoonful. Some patients, for reasons not always known, will refuse food one hour and take it freely the next; it should, therefore, be frequently offered. With milk as a basis, we may add to it, as we are able. Raw egg, gruel, boiled rice, oatmeal, custard, and bread are adjuncts that are nutritious and easily given.

It makes but little difference why patients refuse food, except that a knowledge of the reasons may enable us to overcome their disinclinations. The thing to remember is that they must in some way be made to get enough.

Care of Patients with Insanity, Accompanied by Exhaustion.—There is a condition associated with acute mania or melancholia—though it is sometimes seen in connection with the more chronic forms of insanity,—of exhaustion so overpowering, that it may be rightly compared with the exhaustion of typhoid fever. It may last a few days or a month, or more, if it does not sooner terminate fatally. Instead of the quiet delirium of typhoid fever there is generally violent mania or frenzy. Neither mind nor body is quiet; sleep seems to have fled. The patient may be destructive, constantly out of bed, fighting care, refusing food, and wetting and dirtying himself. With these unfortunate conditions there generally is fever, often to a considerable degree, the heart is feeble, the pulse rapid, the tongue and lips dry and cracked, the teeth covered with sordes, and the body emaciated. Every case does not present all these symptoms, nor show such alarming exhaustion. There are many degrees of severity in this sickness.

Such patients must never be left alone and need constant nursing day and night. They must have food, even if it is given forcibly. They must, if possible, be kept in bed, and covered with clothing, and they must be kept clean. If wakeful, food must be administered during the night, and especially towards morning, which is the time of greatest weakness and physical depression.

Hot baths may be ordered for these patients, and stimulants and medicine to produce sleep left in the care of attendants. How to give the baths and medicine, what results are to be expected, and what dangers are to be feared, will be described later, in the chapter on the administration of medicine.

There are certain symptoms which should warn the attendant of danger, and which often precede death. When any of these are present they should be reported to the physician. They are: partial or complete unconsciousness, slow and labored, rapid, shallow, or irregular breathing, increased weakness and rapidity of heart or pulse, cold hands and feet. Picking at the bedclothes, or at imaginary objects in the air, or vacant staring, are bad symptoms.

The Care of Patients in a Condition of Dementia.—It is to be remembered that dementia may be either, a condition of chronic insanity without recovery, or a less permanent state of mental enfeeblement following the acute attack, and from which recovery may be hoped. In the first of these conditions there is little to be done except to care for the patient. Many are able to do some work, and should be allowed, encouraged, and taught to do it. Others do not know enough to dress, feed, or care for themselves. These must be kept neatly dressed, taken to the table and their food prepared, taken to the bath and closet, taken to walk, and put to bed. If not so attended to, they will degenerate into a ragged, dirty, and even filthy state, and the ward upon which they live will be offensive to the smell. They should be frequently examined for body vermin, as these pests are liable to breed and flourish among these patients. The condition of the demented affords the best evidence of the care given to the patients in an institution. Attendants will often be gratified to see some of these apparently hopeless cases greatly improve and sometimes recover.

If attendants will watch their patients as they come out of acute mania or melancholia and become quiet, they will often notice that they gain in flesh and become demented. The dementia may be but partial, or so very complete that the patient knows nothing. From this they may gradually go on to improvement, or even recovery. They need all the care demanded by the confirmed dement, and, in addition, advantage must be taken of every means to promote recovery. They must be well fed, regularly taken out for exercise, and, as they are able, encouraged to employ themselves. Any symptoms of a return of their more violent condition, any failure to sleep, or change noticed in the health of the patient, should be at once reported.

Care of the Convalescent Patients.—This is the period that precedes recovery from disease. With the insane it is often a critical time, and if not properly cared for they may fail to get well, and become chronic lunatics. The patients, and frequently their friends, think they are well and should be at home. It is the attendant's duty to encourage the patient, and to promote his confidence in the physician. They should not be told of their past conditions, or the disagreeable features of their sickness called to mind, and their last, as well as their first impressions of the asylum should be made pleasant. Sometimes there is a slight return of depression or mania, and the patient may suddenly begin to lose sleep. These conditions must be observed and reported, for it is very easy for patients who are recovering to become as disturbed as when they were first insane, and to suffer a relapse from which they may never recover. It is hardly necessary to remind the attendant that employment, amusement, and all the healthful means of occupation afforded by the asylum, should be judiciously allowed these patients.

Sometimes patients feel too well. They are too contented, happy, and indifferent, and are very active in body and mind. They want to work all day, from early in the morning until late at night. They sing as they work, and talk rather loud and fast. These patients need restriction; they should not be allowed to work too much, so as to overtax their strength. So long, however, as they continue to gain, and sleep well, little is to be feared, and they generally become quieter and recover.

The Care of the Epileptic Insane.—Not all epileptics are insane, but they are all liable to insanity. Generally the most hopeless and difficult to be cared for are brought to the asylum. Epileptics are liable to have fits at any time, but some patients have them at night only. The attack is generally sudden, though sometimes patients have feelings that warn them of their approach. This may precede the fit for a very short time, or the patient may know during the day that he will have a fit during the night.

Epileptic fits are accompanied by convulsions and unconsciousness, and are the type of all convulsions. The unconsciousness may be but momentary, or last an hour or longer, and even prolonged several days; the convulsions may be but the twitching of a few muscles, as of the face, or may consist of the most terrible writhings, and last for several minutes, and be often repeated. Sometimes the fits are ushered in by a scream.

The fit itself is not dangerous to life, but patients may at night turn their face downward and so smother; they may fall from high places, or down stairs, or into the water, or into the fire, and so injure themselves. There is little to do during an epileptic attack. Patients should not be held to prevent the convulsions, but so that they shall not injure themselves. A pillow should be placed under the head and the bands about the neck loosened. The nurse is sometimes given remedies which, if properly administered when the attack is felt to be coming on, may ward off the fit. Nitrite of amyl in small glass pearls is a common remedy. It is to be broken in a handkerchief and several strong breathfuls taken.

At their best, epileptics are cross, irritable, quick-tempered, unreasonable, and quarrelsome, and they will often give a blow at slight, or even for no provocation. After a fit they are frequently dangerous and always require guarded care and watching. As has been said, they may soon recover their natural condition, or remain in a more or less prolonged state of unconsciousness, or they may pass into a condition that appears natural, but in which they have but little or no appreciation of their situation or surroundings, or remember afterwards what they do. In these states they may, without warning, make violent assaults, commit murder or suicide, or set things on fire. Sometimes they do outrageous acts, such as beating their own children to death against the wall, or mutilating them, or roasting them to death on the stove. Many often suffer from hallucinations or illusions of sight or hearing, and have delusions of impending harm or assaults, and think they must defend themselves.

Care of Patients with Paresis.—This is a form of insanity characterized by progressive dementia and increasing bodily enfeeblement and paralysis. The paralysis is partial, not complete ; the patient's walk is feeble, unsteady, and shuffling ; the hands are tremulous, lose their fineness of touch and ability to do work and write ; there is twitching in the muscles of the tongue and about the mouth, and the speech is thick and indistinct. As the disease progresses the patient becomes helpless, bedridden, wet, and filthy. The result is always death. Convulsions like those of epilepsy are liable to occur, from which the patients may rally, or in which they may die or linger a few days. In the earlier stages the patients are often strong, and controlled by delusions and hallucinations that make them violent. Sometimes they are simply good-natured and easily managed. They generally have very exalted and extravagant delusions, and are without appreciation of their condition or surroundings, and are irritated at the control of the asylum, and on account of their unreasonableness they can rarely be allowed the liberty others enjoy.

Paretics often eat ravenously and rapidly, they stuff their mouths full of food and so choke themselves. Their condition of paralysis may render them unconscious of danger and powerless to help themselves. The care needed by bedridden, filthy paretics is practically the same demanded by helpless paralytics, the old, feeble, or demented class, and all others who cannot care for themselves.

Care of the Paralytic, Helpless, Bedridden, and Filthy Patients.—There are many patients in an asylum who are indifferent to all the wants of nature, who wet and dirty themselves. Some of these patients are bedridden; some are about the ward, but demented; some are violent and maniacal, and some from delusions make their persons and rooms as filthy as possible. Much can be done with many of these patients by regularly taking them to the closet, and their bad habits may in this way be broken up. Patients of this class should be visited during the evening, attended to frequently by the night watch, and seen the first thing in the morning. Patients, when dirty, should be thoroughly washed and carefully dried. Their beds should be cleaned and changed, and during the day clean clothing should be given them as often as required.

The greatest danger that comes from not keeping patients clean is the formation of bed-sores.

Bed-Sores.—Bed-sores occur in patients long confined to bed, and who suffer from exhaustive diseases. Paralytics and paretics are particularly liable to them, the diseased condition of the nerves allowing the tissues to break down easily. Sometimes the fingers or toes of a paretic become gangrenous or large surfaces of the skin die, and sometimes deeper tissues slough away rapidly. These conditions may come on in a day or a night.

Patients who are wet and dirty are more liable to have bed-sores. They will always appear in a bedridden paretic in a few days if not kept perfectly clean. They most frequently occur over bony projections where the weight comes in lying, as upon the hips, back, or shoulders.

Such patitents, should, if possible, be made to sit up several hours every day, or placed first on one side, then on the back, and then on the other side. If it can be done, they should, as they lie in bed, rest their hips on an inflated rubber ring, and if the skin is red the part should be bathed in diluted alcohol. After being bathed and dried the skin about the hips should be dusted with some dry powder. Powdered oxide of zinc is perhaps the best, but ordinary corn-starch flour is valuable and serves a good purpose. Insane patients frequently will resist all care and every effort to prevent bed-sores, tearing off the bandages and dressings and picking and irritating the sores.

Bed-sores should never be allowed to come because of want of attention or cleanliness, but there are conditions in which they will appear in spite of every preventive.

Bed-sores once formed should be treated as ulcers and according to the direction of the physician.

CHAPTER IX.

SOME OF THE COMMON ACCIDENTS AMONG THE INSANE, AND THE TREATMENT OF EMERGENCIES.

THE insane, like others, may suffer from almost any accident. It is not intended to treat of all accidents, nor how to care for every emergency. This is so large a subject as to demand a separate text-book, and there are several excellent ones, that attendants would do well to read. But there are among the insane certain kinds of accidents that are likely to occur, certain classes who are liable to receive accidents, and certain emergencies that frequently have to be cared for by the attendant, and these will be described. Every injury received by a patient should be immediately reported to a physician.

Attendants, in the care of the insane should always remember the liability to accident and guard against it. The old, the feeble, the paralytic, and paretic need special care. They are weak, easily pushed over, or stumble and fall, and they cannot break the weight of their fall, or so defend themselves; they are irritable, childish, and often provokingly troublesome to the other patients, and their bones seem to be easily fractured. Some injuries are self-inflicted, some come to the patient in consequence of his own or others' violence, and some, as has been said, from the very weakness of the patient.

Care of Fractured Bones.—Any of the bones may be fractured, and from slight cause. The bones most frequently fractured are : the collar bones, the ribs, the bones of the forearm just above the wrist, the bones of the lower leg and of the thigh. This last bone, the femur, is among old people most frequently broken at its neck, which is the constriction of the bone just below the rounded end that fits into the joint at the hip.

Fractures should, as much as possible, be let alone till the physician comes. The parts should be kept quiet so as not to cause unnecessary pain, and do further injury. By rough handling it is very easy to push a fragment of bone through the skin, thus making a simple fracture a compound one. When a rib is fractured a sharp end may pierce the skin or the lung; either complication is serious. If the lung is injured the sputa will be bloody, and the appearance of such a condition should be at once reported. Sometimes patients are violent after the injury and need to be firmly held, and sometimes they have to be carried to the ward from the outside, or placed upon a bed. Always carry the fractured limb as well as the patient.

If temporary splints are put on do not make them too tight, and loosen them from time to time as needed. The extremities sometimes swell rapidly after a fracture, and the splints may so stop the circulation that, in a few hours, gangrene may be caused by them. Besides, many patients cannot tell us if the part is swollen or painful.

The Care of Wounds.—Bites. Insane patients often bite others and penetrate the skin. They may be very angry, their mouths foul and running with saliva, and this irritating substance introduced into the wound by the teeth may set up an ugly inflammation. The wound should be immediately and thoroughly washed. It should be well cleaned with a wet sponge or cloth, and soaked in warm water. A good after-dressing is powdered iodoform, sprinkled over the wound.

Wounds of the Head.—These wounds are quite common. They should be thoroughly washed and cleaned from dirt and hair. Hemorrhage may be controlled by continued pressure upon the bones of the skull, and if an artery is cut, it can in this way be kept from bleeding till the physician arrives. Most wounds of the head, even though large, generally heal quickly, but the most trifling ones may assume serious proportions, and even prove fatal. If within two or three days heat, pain, redness, and swelling appear, pus is probably forming beneath the scalp, and this, within a few hours, may spread under a large surface and do serious injury, or erysipelas may be set up.

Injuries from Blows on the Head.—Persons are sometimes stunned by blows on the head. They should be placed in bed with the head elevated, and kept perfectly quiet till the doctor comes. Efforts should not be made to arouse them, they should not be given liquor of any kind, but ice may be applied to the head. The danger to be feared is from the skull being fractured, or from bleeding vessels inside of the skull. Either of these conditions may, by pressure upon the brain, cause unconsciousness, paralysis, and death.

The Care of a Cut Throat.—Patients may cut their

throats from ear to ear and do really little injury, or they may make a small stabbing wound and divide a large blood-vessel and die almost immediately, or they may cut the windpipe and not cut the blood-vessels. The windpipe you can notice upon yourselves as a large, stiff tube, prominently situated in the middle and front of the neck; the blood-vessels are together on each side of the windpipe, and situated quite deep down among the muscles, and the carotid artery may be felt beating by the finger. Little can be done by the attendants to stop the flow of blood, even if the great blood-vessels are not cut. The head should be kept bent forward and the chin pressed against the chest.

After the physician has dressed the wound, constant watching day and night may be required to prevent the patient tearing off the bandages and reopening it. This same rule of watchfulness applies to the after-care needed to be given to many cases of fracture, and other serious injuries among the insane.

Care of Wounds of the Extremities with Hemorrhage.—The hemorrhage from most simple wounds involving the cutting of skin and flesh or small arteries, can usually be controlled by direct and continued pressure. This may be done by a pad made of cloth, packed and pressed into the wound, or lint may be used in the same way. Water as hot as can be borne poured into the wound will frequently stop a hemorrhage when other means fail; cold applications and ice are also useful. If dirty, a wound should be thoroughly cleaned, being washed, and, if necessary, soaked in warm water. Iodoform sprinkled so as to cover wounds, is the best dressing for all attendants or nurses to apply, while awaiting directions from a physician. It keeps them clean, promotes healing, and lessens the danger of inflammation or the formation of pus.

When the arteries of the extremities are cut, pressure should be made on the large artery leading to the part. When the wound is high up on the arm, pressure is made by the fingers or a padded key upon the artery that lies back of the collar bone, and the attempt should be made to press it against the bone. This is a difficult thing to do, but nevertheless it should be attempted. When the wound is lower down, pressure is to be made by the fingers on the inner side of the upper arm, at about the middle point and against the bone. The artery runs downward, near the inner border of the biceps muscle, which is the large, bulging muscle of the upper arm, and can, with a little care, be felt beating by the fingers. Patients in breaking glass often cut one or both arteries at the wrist-joint where the pulse is felt. These are large and bleed rapidly, and when cut need the care just described.

When the artery in the leg is wounded, pressure is to be made on the inner side of the thigh, just below the groin. The position of these large arteries, and how to press against the bone, is best learned by instruction and demonstration from a physician, and with a little practice attendants will be able to easily and successfully do the act.

It is very tiresome to continue pressure with the fingers for a long time, and attendants should relieve one another till the physician comes.

The Care of Sprains.—Sprains are a common accident and easily produced. The great end of treatment is to keep the sprained joint quiet. If the ankle or knee is sprained, the patient should be carried to bed. Perhaps the best early treatment, and the one that gives the greatest relief to pain, is to place the joint in a tub of water as hot as can be borne, and keep it hot by pouring in more. The part should be kept in the water until it is parboiled. The skin of some feeble or paralytic patients is easily scalded, and some cannot tell when it is too hot; the water therefore should never be uncomfortable to the hand of the attendant.

Care of Patients Choking.—This is a frequent accident, and in order to know what to do when it occurs, it is necessary to have a knowledge of the air passages of the throat.

We breathe through the mouth and nose. They open into a common passage, the pharynx, which can be seen by looking into the mouth, lying back of the tonsils. Passing downward, it divides by branching into two tubes; one the windpipe, which is in front, behind it is the œsophagus or gullet.

The point of division is just beyond the tongue, and is almost within reach of the forefinger when crowded into the mouth.

The air we breathe passes through the mouth and nose to the pharynx, thence to the lungs by the windpipe. The food we eat passes from the mouth to the pharynx, and thence to the stomach by the œsophagus.

There is at the opening of the windpipe a cover, the epiglottis, which is generally open, but which closes when food is swallowed and helps to keep food from entering. When a substance touches the opening of the windpipe, we instantly cough to expel it.

A person may choke, when the mouth and the pharynx back of it are filled with food ; or when a piece is lodged in the wind-pipe, or a large piece in the œsophagus at the point of division, and which crowds upon the windpipe, or covers the opening. Food gets into the windpipe, by being drawn in by a sudden and unexpected inspiration of air. This may happen while eating or in vomiting solid food. With this accidental exception all breathing stops during the act of swallowing.

Some patients, from paralysis, especially paretics, do not feel food when it is lodged in the throat ; others, from great dementia, may not know when they are choking, and show no emotional signs of distress. Paretics are particularly liable to bolt their food, and cram the mouth and throat full.

The symptoms of choking are immediate, and if no relief is obtained, the sufferer will die in a few minutes. If the patient knows any thing, he will show immediate signs of distress, violent but ineffectual attempts to breathe, and the face quickly becomes a dark blue color, from the accumulation of carbonic acid in the blood.

Immediate effects should be made to remove the obstruction, and continued until the physician arrives, who is to be sent for at once. Whatever is in the mouth and throat can be easily removed by the fingers ; the forefinger should then be crowded down the throat to feel for other obstructions, care being taken not to push a piece of food into the windpipe. If any thing is felt, it can sometimes be pulled out by the fingers, or a hair-pin may be straightened and bent, or a piece of wire, and an effort made to fish it out. When in the gullet and beyond the fingers, it may be pushed into the stomach by a feeding-tube. Artificial respiration may be needed, but attendants must remember it is of no use until the obstruction to breathing is removed.

Marbles, coins, buttons, pieces of pencils, needles, pins, and fish-bones, are frequently swallowed. The physician should be informed at once.

Directions how to Perform Artificial Respiration.—What is to be done must be done quickly ; tight clothing about the neck and chest must be removed, and the mouth should be cleaned of dirt, water, or any obstruction to the flow of air. The body is then laid out flat on the back, covered, if possible, with light warm blankets, and some article should be folded and placed under the shoulders, so as to raise them three or four inches. The mouth must be kept open, and the tongue pulled well forward, as it is liable to fall backwards, and cover the opening of the wind-pipe. One person, kneeling behind the head, should grasp each arm at the elbow, and, draw them steadily around so that the arms will meet above the head. A strong pull should be made upon them, and they should be held a few seconds. These movements elevate the ribs and enlarge the chest and produce an inspiration.

The arms are then to be brought to the side, and pressed strongly against the lower ribs. This last movement drives the air out of the lungs, and makes an expiration. These manipulations should be repeated, slowly and regularly, about sixteen times a minute, and should, when there is the slighest hope of life, be continued at least thirty minutes. The heart should be listened to, in order to hear if it still beats. Warmth, by hot-water bags, bricks, and soapstones should be secured, care being taken not to burn the skin. The limbs may be gently rubbed with warm cloths, though it is not so important as some well-meaning people think. The rubbing should be towards the heart.

As the breathing begins, it should be still aided by the artificial means as long as necessary. When the patient can swallow, teaspoonful doses of brandy or whiskey, to two or three of water, may be given and repeated several times. As soon as possible the patient should be put in a warm bed, and milk and light food given.

Care of Patients when First Burned.—When a patient's clothing is first on fire, dash water over him if near at hand, if not wrap him in a blanket or some heavy woollen garment, and smother the fire. Then unroll the patient and extinguish the smouldering pieces of clothing. The clothing must be cut and clipped off. Great care must be taken not to tear open the blisters. If any application is made, it may be by linen cloths soaked in sweet or castor oil, or equal parts of linseed oil and lime-water, or a layer of flour and molasses may be applied over the burned surface. These bland substances act largely by excluding the air, which, if blowing ever so quietly, is always painful and irritating, and they also protect the wound from the irritation of the bed and body clothing. Burns from scalding are practically treated in the same way as burns from fire.

Care of Frost-bites.—Toes, fingers, ears, and noses are most frequently frozen. They will sometimes freeze in a few minutes on a very cold day. After a part is frozen there is no feeling of cold or pain, and it looks perfectly white, and is so stiff it may be broken.

Persons who are frost-bitten should not be taken into a warm room. They should be left in a cool room, and the frozen part rubbed with cold water, or ice, or snow. As these last melt they melt the frozen flesh. If the parts are thawed too quickly gangrene is liable to follow.

Care of Patients in States of Unconsciousness.—This is not an accident, but a frequent emergency. The medical word for unconsciousness is *coma*. It may be partial or complete, may come on suddenly or slowly, or may be accompanied by convulsions or paralysis. The more frequent causes of coma, are epilepsy, the convulsions of paresis, blows on the head, hemorrhage in the brain or apoplexy, some diseases of the brain, sunstroke, and some poisons.

When coma comes on, attendants should observe, if it is slow or sudden; if the patient complains of pain in the head; if the respirations are changed, and how; the condition of the pupils, whether large, contracted, uneven, or changeable; if the mouth and face are drawn to one side; if there is any paralysis of the arms or legs; if there are any convulsions, or twitching of muscles; if the patient can be aroused, and from time to time observe and count the pulse.

Apoplexy is a term that is much used, and is a condition of coma, caused by pressure on the brain. This organ is in a tight, rigid box, the skull. If the fluid of the brain is much increased, or blood-vessels ruptured, pressure is the result, and the soft tissues yield, rather than the bony covering. This pressure may destroy or injure the cells and fibres, and so interfere with the function of the part. Another way that apoplexy occurs is by plugging of an artery of the brain, so that it cannot deliver blood to the part to which it goes, and consequently the part loses its ability to perform its function. The plugging is most frequently due to a small clot floating in the blood, and which is usually formed in the heart.

Paralysis and apoplexy are often, through ignorance, used synonymously, but they really mean very different conditions. Paralysis is a loss of power of contracting a muscle, due to disease or injury of the nervous system ; it frequently follows or is associated with apoplexy.

In the case of apoplexy, and most conditions of coma, there is generally little for the attendant to do. The patient should be put to bed, with light coverings, and the head raised on pillows. Do not annoy the patient by trying to rouse him, and do not give stimulants.

Care of Sunstroke.—A sunstroke is a very serious condition, and when it occurs, requires immediate efforts to save the life of the one suffering from it. It generally comes on suddenly, the patient first complaining of the head ; he soon becomes unconscious, the skin hot and dry, and the pulse full and bounding. The treatment consists of taking the patient to a cool, shaded place, removing all unnecessary clothing, applying ice or cold water to the head, and bathing or sponging the body in cold water. If the patient recovers, the temperature will fall under this treatment. If the heart begins to fail, or the pulse becomes weak or fluttering, small doses of whiskey and water may be given and repeated.

Patients should not be taken out in the fields nor exposed places on very hot days, except as ordered by the physicians; they should wear light clothing and a straw hat; if permitted to go out, they should not overwork, and should be allowed frequently to rest in the shade. Patients are easily injured by working in the sun; headache caused, recovery retarded, and bad symptoms brought back, without having the alarming conditions of sunstroke.

Unconsciousness from Poisoning.—Opium and its preparations, including morphine, chloral, and the two extracts of hyoscyamus, now so much employed in asylums, namely, hyoscine and hyoscyamine, are medicines frequently given, that poison in over-doses and produce coma.

These medicines and their effects will be described in the next chapter, and at the same time the symptoms of poisoning by them, and the treatment.

Poisoning.—Poisonous drugs are not kept upon the wards. Attendants frequently have strong ammonia in their rooms to clean their clothing, and a patient may get it and drink it. It is a strong alkali, and burns the throat and mouth. Vinegar is the best ready antidote, but should be given immediately or not at all. Soft soap is a strong alkali, and if eaten becomes an irritating poison. Again vinegar is the best antidote.

The best antidotes for acids are soda, lime-water, soap-suds, and chalk; for alkalies, weak acids, such as lemons, oranges, vinegar, or cider. Olive oil, eggs, and mucilaginous drinks are the most bland and soothing remedies to give. To vomit a person who has taken poison, give a pint or a quart of lukewarm water; to it may be added one or two teaspoonfuls of mustard. Syrup of ipecac is a common remedy, the dose is a teaspoonful, and repeated in ten minutes if necessary. It assists vomiting to tickle the throat with a finger or a feather. If after poisoning there is depression or approaching coma, very strong tea or coffee is the best stimulant, and it is as well an antidote to many poisons. If the heart and pulse are very weak, whiskey diluted with water may be given and repeated.

Injury from Eating Glass.—Patients sometimes eat glass. This injures by the edges cutting and inflaming the walls of the stomach and intestines. This may be so severe as to cause death. In the treatment do not give an emetic or a cathartic. Such food as has a tendency to constipate the bowels, and such as will also enclose the glass and coat its sharp edges, is to be given. Potatoes, especially sweet, oatmeal, or thick indian-meal pudding, are appropriate. Cotton, which is generally at hand, will, if swallowed, engage the glass in its fibres, and so protect from injury.

Injury with Needles.—This is a self-injury, but it may be severe and require immediate attention. Patients may open a vein or an artery with a needle, or plunge it into the eye. But the more common way is for a patient to stick many needles under the skin, sometimes to the number of several hundred. Sometimes patients introduce them near the heart or lungs, and as a needle will often "travel" when in the flesh, it may work its way into a deeper part, and so a number get into the lungs or the heart, causing death. Within a few weeks I saw two needles taken from a man's heart, who died in consequence of their presence there. An attempt or desire to so injure one's self should be guarded against by the attendants, and if accomplished should be at once reported to the physician, that efforts may be made to extract the needle.

CHAPTER X.

SOME SERVICES FREQUENTLY DEMANDED OF ATTENDANTS, AND HOW TO DO THEM.

THE Administration and Effect of Medicine.—The only proper way of giving medicine is by using standard weights and measures. Dropping medicine, or using spoons or cups, is not sufficiently accurate. A drop may be half a minim, or as large as two or even three. The modern teaspoon holds ninety or more minims, and a tablespoon more than half an ounce.

Medicines are introduced into the system through the stomach, the lungs, the rectum, the skin, or by being injected into the tissues, under the skin. They are either local or general in their effects. A blister or a poultice is a local remedy, so is an emetic, that acts by irritating the walls of the stomach. General medicines are absorbed into the blood, and carried to different parts of the body.

The following are a few of the reasons for which medicine is given: to relieve pain, to give sleep, to produce vomiting, to check vomiting, to move the bowels, to check diarrhœa, to assist digestion, to produce a greater or diminished flow of urine, to increase the perspiration, to increase the red blood corpuscles, to check hemorrhage, to regulate the action of the heart, to overcome the effects of poison, to increase or diminish the amount of blood in the brain, to control spasm, to diminish the temperature in fever.

In some cases the effect desired and expected from a medicine given to a patient is told to the attendant, who should closely observe and be able to report the result. Sometimes medicines are left in the hand of the attendant, to give in repeated doses, at stated intervals, till a desired effect is produced. The attendant is also instructed to watch for certain symptoms which show that the medicine is doing harm, when it is to be discontinued. An attendant, who has studied and learned, "how and what to observe" in his patient, will be able to give intelligently any medicine ordered by a physician.

Sometimes medicines, given in large or long-continued doses, cause symptoms that an attendant should notice and report to the physician; some of these are, eruptions on the face and body, puffiness about the eyes, irritation and running of the eyes, a metallic taste in the mouth, bleeding of the gums or soreness of the teeth and profuse flow of saliva, nausea, vomiting, diarrhœa, constipation, indigestion, ringing of the ears, feeling of fulness in the head, headache, dizziness, drowsiness, coma, convulsions, or convulsive movements of muscles.

In asylums, medicines are mostly sent to the wards in single doses, each cup or bottle being marked with the name of the patient for whom it is intended. The tray in which they are carried should never be set down and left, for a mischievous or suicidal patient may poison himself by taking every thing he can get hold of.

No patient, unless ordered by the physician, should be

allowed to keep his cup and take his medicine at his leisure. Suicidal patients often ask to do this, and then save the medicine, until they have enough to poison themselves. Others will throw the medicine away. The way to administer medicine to the insane is to give it personally to the patient, and also see that it is swallowed. It is a frequent custom of many patients to retain the medicine in the mouth, and, when the attendant has left, to spit it out.

It is often very important that patients should take the medicine ordered, and every effort should be made to induce them to take it. Such patients should be designated by the physician. Night medicines, or those given about bedtime, are usually of great importance. All patients who refuse to take their medicine should be reported to the physician.

The reasons for refusing medicines are various; some say they are perfectly well and need no medical treatment, others think the medicine injures them, that it turns their skin black, or poisons them, or that it is wrong to take it, or displeasing to God; ideas much like those that we learned were the causes for the refusal of food. Attendants are to use every effort to get patients to take medicine, and may employ as much force as they were instructed to use in giving food, but no more.

Patients should not be deceived about medicines, nor told by attendants that it is nothing, that it is only a little water, or some nice drink that is sent to them, nor should an attempt be made to give them, by trying to disguise them in food or drink, except by the permission of a physician. Patients should, on the other hand, be told that it is medicine, that the doctor ordered it for them, that it is for their good to take it, that it is given to help them get well.

The giving of medicine and food is among the most important and frequent duty that an attendant is called upon to perform, or assist others in doing. Attendants must remember that many medicines are injurious or even poisonous, if not properly given, or if mixed with other medicines, or if given to the wrong patient ; they should therefore, never make a mistake, or, if by carelessness they commit one, should immediately report it.

Opium and Some of its Preparations.—Opium is a medicine that is very frequently given to patients in an asylum. The ordinary dose is one grain. *Tincture of opium, or laudanum,* is opium dissolved in alcohol. Ten minims equal one grain of opium. *Camphorated tincture of opium, or Paregoric,* is a weaker alcoholic solution, with some camphor, and flavored with a pleasant aromatic. One half a fluid ounce equals a grain of opium. *Morphine* is a white powder extracted from opium. An eighth of a grain about equals a grain of opium.

Opium, in some of its forms, is a common household remedy. To an adult, not more than one grain should be given ; it should not be repeated more than once, nor less than six hours after the first dose. It would be better if never given, except by a physician's order. Under no circumstances should any one but a physician give it to a weak or old person, or to a young child.

Opium, is given in ordinary doses to relieve pain, to check diarrhœa to relax spasm of muscles, and to produce sleep. The sleep from opium is generally quiet and refreshing, and one from which the patient can be easily aroused.

An attendant will frequently be told when the medicine is given and directed to note and report its effect.

Opium Poisoning.—The taking of opium is a frequent way of committing suicide by persons outside of asylums. Sometimes patients manage to save their doses, or they steal it from the tray, or, if there is some sent to the ward for repeated doses, they secure it through the carelessness of an attendant, or occasionally it is secretly sent to patients by officious outside friends,—thus poisoning by opium sometimes occurs among asylum patients.

The full symptoms of poisoning are profound coma, pupils contracted to pin-points, and which do not respond to light; very slow respiration, often not more than four or six times a minute, but heavy and labored. Sometimes the effect of the drug is but partial, the patient can be aroused for a moment, but falls to sleep again, or the symptoms may be even less pronounced.

The treatment of opium poisoning, before the physician comes, consists in giving *very* strong coffee, or tea, an emetic, and in trying to keep the patient awake by walking him about, or, if this is not possible, to keep him from falling into deeper coma, by shaking, calling loudly in the ear, and striking and slapping the body with wet towels.

Chloral.—This is a white crystal, with a pungent, burning taste. It is always dispensed, dissolved in water, and should be further diluted when given to a patient. The dose is from ten to thirty grains. It is too powerful a drug to be given, except upon the order of a physician.

Chloral is given to produce sleep, which is usually quiet and natural. The effect lasts about four or six hours.

The symptoms of poisoning are not so marked as to make it easy to know that they are caused by chloral. There is generally a weak heart and pulse, and feeble respiration, and the patient is in a deep sleep, from which he may be aroused; or the coma may be profound, and continue uninterrupted till death.

The treatment consists in giving an emetic, stimulants, coffee, and, if necessary, performing artificial respiration.

Hyoscyamine and Hyoscine.—These are extracts, from the leaves and seeds, of the plant hyoscyamus.

These are very powerful medicines, and are never given except on the order of a physician. They are always given in solution.

The action of both is practically the same. In ordinary doses they quiet restlessness, produce muscular weakness, flushing of the face, dryness of the tongue, wide dilatation of the pupils, and frequently cause sleep. These effects should be noticed and reported. These medicines are mostly given to patients who are continually restless, violent, and sleepless, and the object is to bring quiet, repose, and sleep. Large doses may produce coma, very heavy breathing, and great muscular weakness; the pulse however is full and strong, but if it should fail, the physician should be at once sent for.

Alcohol and Stimulants.—It is the alcohol in liquors that intoxicate, and it is that part, also, of liquor that stimulates when given as a medicine. Whiskey, brandy, and gin are about one half alcohol. The dose is a tablespoonful, in water, and not repeated oftener than two or three times. Wines are about one fifth alcohol, beers and cider about one twentieth.

Liquors containing alcohol are never to be given to patients as a beverage, but always as a medicine, and, except in emergencies, never without a physician's order. Do not give them in emergencies, without a good reason for so doing, and not simply because you feel you must do something, for in some emergencies they may do a great deal of harm, and perhaps, a fatal injury.

Alcohol is mostly given to stimulate the action of the heart. A stimulant is something "that arouses or excites to action." It is given (in the doses just mentioned) in accidents, when the heart is very weak, the pulse almost or quite imperceptible, the face pale and pinched, and the extremities cold.

In continued sickness, with exhaustion, stimulants are sometimes left with the attendant to give, with directions about the size of the dose and its frequency. If it quiet the patient, strengthen the heart and pulse, it is doing good; but if restlessness comes on, the face becomes flushed, or if the pulse is made more rapid and feeble, it is probably doing harm, and should be discontinued, and the physician informed.

Dry and Moist Heat.—In applying heat, either dry or moist, to the insane, care must always be used to protect the skin from being blistered. This happens very easily when it is applied directly to old, feeble, paralyzed, or paretic patients, and also to those who are too demented to complain if they are being burned. Burns are very serious accidents among this class of patients, and may, if they extend over a large surface, even though not deep, heal with difficulty, and even prove fatal.

Dry heat is applied by means of rubber bags filled with hot water, hot-sand bags, bricks, or soapstones, and by the lamp bath. Moist heat by hot baths, fomentations, turpentine stupes, and poultices.

Hot Baths and Wet Packing.—Hot baths are sometimes prescribed for patients. The water should be about 100 degrees F., and, if ordered, slowly increased to 110°. The patient is to be left in as long as directed, which may be but a few minutes, or half an hour, or even longer. Sometimes a blanket is ordered thrown over the tub, the head only being uncovered.

When the bath is being given, the pulse should be counted ; if it become weak and rapid, if the face become flushed, and the patient complains of dizziness, or if the lips show venous congestion, the patient should be at once removed, and, unless there is immediate recovery from these evil effects, the physician should be informed.

In giving a wet pack, the patient is wrapped in a sheet, without any clothing, wet either in cold or warm water, as ordered, and then rolled in a blanket, put to bed, and left in it as long as directed.

These methods of treatment are frequently ordered by physicians for patients who are restless, violent, and sleepless, with a view of giving quiet and sleep. The attendant should observe and report the result.

Application of Cold.—The attendants are frequently ordered to apply ice to some part of the body, for the purpose of producing local cold. The ice should be broken into small pieces and put into a bladder, or rubber bag, partly filling it. It remains sufficiently cold until all the ice is melted.

Another way is to put a piece of ice in a sponge and bathe the part. When cold cloths or compresses are applied, the heat of the body soon warms them, when they become warm applications and act as a poultice ; they should therefore be frequently changed. In applying moist dressings care must be used not to have any leaking nor wetting of the bed or clothing.

Hypodermic Injections.—Morphine, hyoscyamine, or hyoscine, in solution, are frequently injected under the skin. The direction to do this, and the quantity to be given, will, in every case, be ordered by the physician. A fold of the skin is held between the finger and thumb, while the needle held in the other hand is quickly pushed straight under the skin to the depth of about half an inch. Care should be used to inject no air, and not to inject the contents of the syringe, into a vein.

Forcible Feeding with the Stomach-Tube.—Attendants are frequently called upon to assist in the forcible feeding of patients, and in some cases may themselves be directed to do it. The dangers of feeding are that the pharynx may be filled with fluid, and the patient choke, or it may be drawn into the lungs, that the wedge with which the mouth is held open may be so loosely held that in the struggle of the patient the soft parts of the mouth may be injured, and occasionally it happens that the mere pressure of the tube causes choking.

Attendants should watch the process of feeding, and particularly the face, for symptoms of venous congestion, and report to the physician any thing they see that denotes danger.

In preparing for feeding, attendants must see that the food is properly made ready. If any thing is to be mixed with milk, it should be mixed so as to be perfectly smooth, without lumps, and so it will run easily through the tube. If some concentrated food is used, it is better to put it in a small quantity of milk, just enough to make it liquid, that it may be given first. Medicines ordered for feeding are not to be mixed with a large quantity of milk, but saved, that they may be given directly from the dispensing bottle whenever the physician desires to do so.

Every thing should be got ready for feeding before the physician arrives. Upon a tray should be all the feeding apparatus—the food and medicine, several spoons, and cups, and a pitcher. Near at hand should be plenty of water, soap, and towels, and a tin basin. It is very provoking to have to wait for many things to be brought after the patient has been got ready.

Many patients are easily fed. Some like it, but some violently and furiously resist. Such patients should be restrained to a chair fixed to the floor, and the more securely this is done the more easily can they be fed, and with less fatigue and danger of their being injured.

The patient's clothing should be well protected from being soiled, by towels about the neck, and a basin should always be held under the chin to catch falling liquids and any thing vomited. The holding the head and wedge is an important matter, and is some thing that belongs to the attendant to do. The attendant stands behind the patient, and holds the chin by the right hand, and with the left firmly grasps the wedge, which is inserted straight into the mouth, between the back teeth, about two or three inches. The wedge should be grasped with the palm upwards, and the little finger and side of the hand should be pressed firmly against the chin. If held in this way there is little danger that in violent struggles, the wedge can be suddenly driven backward and wound and tear the soft parts of the mouth. If the throat fills with fluid, the attendant who holds the head should bend it far forward, that it may, if possible, run out of the mouth.

After feeding, patients' faces should be washed. They should be watched for some time to see that they do not vomit, or, as is often the case, that they do not make themselves vomit.

There is no special difference in caring for a patient fed with a nasal tube, except that the wedge is not used.

If attendants are allowed to feed, they must remember all the dangers, and guard against them. In introducing the tube, the forefinger of the right hand is to be introduced at the same time, and, as the tube passes over the tongue it is to be turned downward by the finger and *gently* pushed into the œsophagus. If there seem to be unusual difficulty in so doing, severe and unusual struggling, or the slightest symptom of danger, cease the effort to feed, and report to the physician.

Of course no attendant would undertake to feed any patient unless ordered to do so by the physician, which order would be given, if at all, only after careful training and in cases easily fed.

Nutritive Enemata.—It is often necessary to feed patients by the rectum. This is done by injecting food, to the amount of four or six ounces. Care should be used to inject no air. The nozzle of the syringe well oiled is to be gently introduced, and the fluid slowly forced into the bowel. The patient should lie on the left side, near the edge of the bed, with the knees well drawn up. If the patient resist, he must be placed upon the back, the legs separated and firmly held. This may require four or six attendants, but enough should always be at hand to thoroughly and easily overcome the patient. Before giving the first injection of food the bowels should be moved by an injection of soap and water. Sometimes the injected food escapes from the rectum. The patient should be watched to see if this happens. In such a case a long tube can be introduced into the rectum, about four or six inches, and the food injected through it. The tube should be well oiled, and introduced slowly and with gentle force.

Patients often thrive upon this way of feeding. The character of the food will be ordered by the physician.

STUDENTS' MANUALS.

Manual of Prescription Writing. By MATTHEW D. MANN, M.D., late Examiner in Materia Medica and Therapeutics in the College of Physicians and Surgeons, New York. Revised edition. 16mo, cloth. $1.00.

Manual of Practical Normal Histology. By T. MITCHELL PRUDDEN, M.D., Director of the Physiologica, and Pathological Laboratory of the Alumni Association of the College of Physicians and Surgeons, N. Y., etc. 16mo, cloth. $1.25.

Students' Manual of Venereal Diseases, being the University Lectures delivered at Charity Hospital, B. I., during the Winter Session of 1879-80. By F. S. STURGIS, M.D., Clinical Lecturer on Venereal Diseases in the Medical Department of the University of the City of New York, etc., etc. Fourth edition. 16mo, cloth. $1.25.

Students' Manual of Diseases of the Skin. By L. D. BULKLEY, M.D. Large 16mo. $1.25.

Students' Manual of the Diseases of the Nose and Throat. By J. M. W. KITCHEN, M.D. 16mo, illustrated, cloth. $1.00.

Students' Manual of the Pharmacopœia of the Diseases of the Throat. By GEORGE M. LEFFERTS, M.D. $1.00.

Students' Manual of Rational Electro-Therapeutics. By R. W. AMIDON, A.M., M.D., Lecturer on Therapeutics at the Woman's Medical College of the N. Y. Infirmary, etc., etc. 16mo. $1.00.

Students' Manual of Diseases of the Nerves. By E. C. SEGUIN, M.D. (*In preparation.*)

G. P. PUTNAM'S SONS, NEW YORK AND LONDON.

SUGGESTIVE THERAPEUTICS. A Treatise on the Nature and Uses of Hypnotism. By H. BERNHEIM, M.D., Professor in the Faculty of Medicine at Nancy. Translated from the second and revised French edition, by CHRISTIAN A. HERTER, M.D., of New York. Octavo, cloth $3.50

"I present this volume to the English-speaking medical public in the belief that it throws important light upon a subject which has too long been misunderstood and ignored."—EXTRACT FROM TRANSLATOR'S PREFACE.

PSYCHIATRY. A Clinical Treatise on Diseases of the Fore-Brain, Based upon a Study of its Structure, Functions, and Nutrition. By THEODOR MEYNERT, M.D., Professor of Nervous Diseases and Chief of the Psychiatrical Clinic in Vienna. Translated (under authority of the author) by B. SACHS, M.D. Octavo, cloth $2.75

"We most earnestly urge our readers to put this work in their libraries as one that will prove indispensable."—*Quarterly Journal of Inebriety*, Jan., 1886.

THE INSANE IN FOREIGN COUNTRIES. By WILLIAM P. LETCHWORTH, President of the New York State Board of Charities. Octavo, cloth . . $3.00

THE ERRORS OF REFRACTION. By FRANCIS VALK, M.D., New York. 245 pages. Numerous illustrations (some in color). $3.00

PHYSIOLOGICAL NOTES ON PRIMARY EDUCATION AND THE STUDY OF LANGUAGE. By MARY PUTNAM JACOBI, M.D. 12mo, cloth . $1.00

ESSENTIALS OF PHYSICS AND CHEMISTRY. By CONDICT W. CUTLER, M.S., M.D. Third edition, enlarged and revised. Cloth $2.00

THE STORY OF THE BACTERIA. By T. M. PRUDDEN, M.D., author of "A Manual of Practical Normal Histology." 16mo, cloth 75

HYSTERIA AND OTHER NERVOUS AFFECTIONS. A Series of Essays, by MARY PUTNAM JACOBI, M.D. $2.00

CONTENTS—1. Loss of Nouns in Aphasia. 2. Case of Nocturnal Rotary Spasm. 3. The Prophylaxis of Insanity. 4. Antagonism between Medicines, and between Remedies and Disease. 5. Hysterical Locomotor Ataxia. 6. Consideration on Tumors of the Brain.

PUBLICATIONS OF G. P. PUTNAM'S SONS.

ALT. The Human Eye in its Normal and Pathological Conditions. By ADOPH ALT, M.D., Lecturer on Ophthalmology in Trinity Medical College, Toronto, with the editorial assistance of T. R. POOLEY, M.D. 8vo, illustrated. $3.00.

ALTHAUS. On Sclerosis of the Spinal Cord. Including Locomotor Ataxy, Spastic Spinal Paralysis, and other System Diseases of the Spinal Cord; their Pathology, Symptoms, Diagnosis, and Treatment. By JULIUS ALTHAUS, M.D. With nine illustrations. 8vo, cloth. $2.75.

BROWNE AND BEHNKE. Voice, Song, and Speech. A Practical Guide for Singers and Speakers, from the Combined View of the Vocal Surgeon and the Voice-Trainer. By LENNOX BROWNE, F.R.C.S., Surgeon to the Throat and Ear Hospital, London, and EMIL BEHNKE, author of "The Mechanism of the Human Voice," With numerous illustrations by wood-cutting and photography. 8vo, cloth. $4.50.

BUHL. Inflammation of the Lungs, Tuberculosis, and Consumption. By Professor LUDWIG BUHL, of Munich. Translated by Drs. M. D. MANN and S. B. ST. JOHN. 8vo, cloth. $1.50.

BULKLEY. Eczema and its Management. A Practical Treatise Based on the Analysis of Two Thousand Five Hundred Cases of the Disease. New and Revised Edition. By L. D. BULKLEY, M.D. Large 8vo. $3.00.

—— Acne and its Treatment. A Practical Treatise Based on the Study of One Thousand Five Hundred Cases of Diseases of the Sebaceous Glands. 8vo, illustrated. $2.00.

CLARKE. A Manual of the Practice of Surgery. By FAIRLIE CLARKE, M.D., F.R.C.S., late Assistant Surgeon to Charing Cross Hospital. Third Edition, Revised, Enlarged, and Illustrated by 190 Engravings on Wood. $2.50.

CORNING. Brain-Rest. By J. LEONARD CORNING, M.D. 16mo, cloth. $1.00.

CROOM. Manual of the Minor Gynecological Operations and Appliances. By J. HALLIDAY CROOM, Physician to the Royal Maternity Hospital, Edinburgh. Second Edition, Revised and Enlarged, with 12 Plates and 40 Wood-cuts. $2.25.

CUTTER. A Dictionary of the German Terms Used in Medicine. By GEORGE R. CUTTER, M.D., Surgeon of the N. Y. Eye and Ear Infirmary, etc., etc. 8vo, cloth extra. $3.00.

G. P. PUTNAM'S SONS, NEW YORK AND LONDON.

DARLING & RANNEY. **The Essentials of Anatomy,** prepared as a Text-book for Students, and a Work of Easy Reference for the General Practitioner. By WM. DARLING, Professor of Anatomy, and A. L. RANNEY, Adjunct Professor of Anatomy, in the Medical Department of the New York University. 8vo. $3.00.

RANNEY (A. L., *Editor*). **Anatomical Plates.** Arranged as a companion volume for "The Essentials of Anatomy," and for all works upon Descriptive Anatomy. Comprising 439 designs on steel by Prof. J. N. MASSE, of Paris, and numerous diagrammatic cuts selected or designed by the Editor, together with explanatory letter-press. Large 8vo, cloth extra. $3.00.

DOWSE. **Syphilis of the Brain and Spinal Cord.** Showing the part which this agent plays in its production of Paralysis, Epilepsy, Insanity, Headache, Neuralgia, Hysteria, Hypochondriasis, and other Mental and Nervous derangements. By THOMAS STRETCH DOWSE, M.D., Fellow of the Royal College of Physicians in Edinburgh, President of the North London Medical Society, etc., etc. 8vo, illustrated. $3.00.

—— **Neuralgia: Its Nature and Curative Treatment.** Forming Part II. of "Diseases of the Brain and Nervous System. 8vo, cloth extra. $2.25.

—— **The Brain and the Nerves.** 8vo, cloth. $1.50.

JACOBI. **Infant Diet.** By A. JACOBI, M.D., Clinical Professor of Diseases of Children, College of Physicians and Surgeons, New York. Revised, enlarged, and adapted for popular use by MARY PUTNAM JACOBI, M.D. 12mo, boards. 50 cents.

KITCHEN. **Consumption: Its Nature, Causes, Prevention, and Cure.** By J. M. W. KITCHEN, M.D. 12mo, cloth. $1.25.

KNAPP. **Cocaine and its Use in Ophthalmic and General Surgery.** By H. KNAPP, M.D. With supplementary contributions by Drs. F. H. BOSWORTH, R. J. HALL, E. L. KEYES, and WM. M. POLK. 8vo, cloth. 75 cents.

MATTISON. **The Treatment of Opium Addiction.** By J. B. MATTISON, M.D. 8vo, cloth. 50 cents.

MITTENDORF. **A Manual of Diseases of the Eye and Ear.** By W. F. MITTENDORF, M.D. Third Edition, Revised. Fully Illustrated. 8vo, cloth extra. $4.00.

PARKER. **Cancer: Its Nature and Etiology.** With Tables of 397 Illustrated Cases. By WILLARD PARKER, M.D. 8vo, cloth. $1.50.

G. P. PUTNAM'S SONS, NEW YORK AND LONDON